Iraq and the Politics of Oil

IRAQ AND THE POLITICS OF OIL

An Insider's Perspective

Gary Vogler

University Press of Kansas

Published by the University Press of Kansas (Lawrence, Kansas 66045), which was organized by the Kansas Board of Regents and is operated and funded by Emporia State University, Fort Hays State University, Kansas State University, Pittsburg State University, the University of Kansas, and Wichita State University.

Library of Congress Cataloging-in-Publication Data
Names: Vogler, Gary, author.
Title: Iraq and the politics of oil : an insider's perspective / Gary Vogler.
Description: Lawrence, Kansas : University Press of Kansas, 2017. | Includes bibliographical references and index.
Identifiers: LCCN 2017026728| ISBN 9780700625062 (cloth : alk. paper) | ISBN 9780700625079 (ebook)
Subjects: LCSH: Petroleum industry and trade—Political aspects—Iraq. |United States—Foreign economic relations—Iraq. | Iraq—Foreign economic relations—United States. | Office for Reconstruction and Humanitarian Assistance. | Coalition Provisional Authority. | Postwar reconstruction—Iraq.
Classification: LCC HD9576.I72 V64 2017 | DDC 338.2/72909567090511—dc23
LC record available at https://lccn.loc.gov/2017026728.

British Library Cataloguing-in-Publication Data is available.

Printed in the United States of America

10 9 8 7 6 5 4 3 2 1

*To Barb, my bride of forty-four years—
you were the rock, my foundation, that enabled me
to realize my Iraq adventure. The visits home
during my seventy-five months of deployment
always ended with very difficult farewells,
but your strength supported me. Thank you.*

Contents

Illustrations

Maps

Figures and Tables

Photographs

Preface

This book is more than just my memoirs. It covers the Iraq oil story over my five months of prewar planning and seventy-five months in Iraq—October 2002 until September 2011. As a member of the Energy Infrastructure Planning Group (EIPG), I participated in the interagency oil sector planning at the Pentagon. After EIPG, I moved to retired Lieutenant General Jay Garner's Office for Reconstruction and Humanitarian Assistance (ORHA) team as a senior oil advisor in January 2003. ORHA started planning at the Pentagon and then deployed to Kuwait on March 17, 2003. We began moving into Iraq about a month later. As the first American civilian to enter the Ministry of Oil headquarters in Baghdad, on April 25, 2003, my initial task was to appoint the first post-Saddam Iraqi oil sector leadership team on May 3, 2003. The Coalition Provisional Authority (CPA) under Ambassador L. Paul Bremer governed Iraq until June 2004. My tenure as an oil advisor continued through the CPA era.

The book addresses the important role that oil played after CPA. It includes oil's role in northern Iraq during the surge of 2007, which involved the role of the Bayji Refinery and distribution of its petroleum products throughout the Sunni north. It covers the role of the Energy Fusion Cell in 2007 and 2008 to focus the oil and electricity sectors in support of the surge security objectives. It covers the increased oil production after 2009 and the role that the coalition forces played to support and push the export infrastructure project. This led to record production levels at the end of 2015, moving Iraq into the number-two production position in OPEC, behind Saudi Arabia.

Iraq's oil production increased from 2.5 million barrels a day in early 2003 to 4.4 million barrels a day at the end of 2015, an increase that any oil company would envy. This book covers this thirteen-year period from the perspective of the US government and the coalition forces in a way that has never been told. Stories and vignettes are used throughout the book to highlight how the coalition forces partnered with the Iraqis in the Ministry of Oil to achieve successes in the oil sector. It also tells the stories of how the US government imposed obstacles when arrogance hindered us from partnering in 2003. Fortunately, the successes outnumbered the failures. The oil story highlights the successes of many of the participants in the oil sector, including Iraqis, coalition military, and US government civilians.

It was always my intention to write a book about my Iraq oil adventure after returning from Baghdad in 2011. My family never understood what the attraction could be that took me to Iraq for seventy-five months between 2003 and 2011. They deserved an explanation. They heard some of my daily experiences while in Baghdad, but nothing that required more than five minutes of explanation on a long-distance phone call. My original reason for writing this book was to document my journey for my family. It quickly expanded to the broader purpose of telling the oil story to a larger audience. It provides historical information for the global oil industry and a historical military account of Iraq oil planning and execution. The book is used to thank the many Americans and Iraqis that enriched my life; the reader will note that I try to mention the name of each individual. Finally, the book provides a historical reference for anyone interested in the facts of our government's Iraq oil sector involvement.

My wife did the initial editing of the manuscript for me. This process gave her a better appreciation of my Iraq attraction. She sacrificed a lot during my absence, much more than me. I loved the work. It was the most rewarding work in my life.

My sense of urgency in writing this book increased after my participation in a Rachel Maddow documentary. Maddow's producers contacted me a couple of times in 2013 to convince me to agree to an interview, but I turned them down. Ambassador L. Paul Bremer contacted me and asked that I join him and Meghan O'Sullivan to help refute the suspected Maddow documentary theme: that the Iraq War was all about oil. I agreed to join them.

My preparation for the documentary made me realize that the Iraq oil story had never been written. My research while preparing for the interview made me realize something significant: there were Iraq oil events that should be more thoroughly investigated and documented. Some of the conclusions from my in-depth research were a surprise, even to me. There was an oil agenda that may have had a significant influence on the decision to go to war. Before 2014, I refuted any comment about oil being a significant motivator for our invasion of Iraq. My opinion has changed. There was an oil agenda, but it was not the oil agenda that most people believe. As a secondary objective of the book, this oil agenda is described; it is something that currently troubles me. What role did it play in the decision for war? It was an agenda to provide Iraqi oil to Israel. Pieces of this agenda are examined throughout the book.

Several of my ORHA and CPA colleagues told me that the oil story should be told and that I needed to tell it. No one else participated more in both the planning at the Pentagon and the subsequent execution of that plan in Iraq. My second tour started in 2006 and included oil reconstruction until late 2011. All of that time in Iraq was under the auspices of the Department of Defense (DOD). My two deployments were as a government civilian appointee or independent contractor. I am not aware of anyone under the DOD who spent more time in Iraq between 2003 and 2011. My service was under retired Lieutenant General Jay Garner, Ambassador Bremer, General Casey, General Petraeus, General Odierno, and General Austin. The ambassadors during my second tour included Khalilzad, Crocker, Hill, and Jeffrey.

One final trip was made in May and June 2015 under the State Department. My engagement was to lead a two-month USAID project that was an oil assessment for the new oil minister, Adil Abdul-Mahdi al-Muntafiki, and Ambassador Stuart Jones. The epilogue discusses the issues facing the oil sector at the end of 2015 and why the large production increases of 2012 through 2015 should moderate over the next five years.

It was a pleasure to relive my Iraq oil adventure while writing this book. I hope you enjoy reading the book as much as I enjoyed writing it.

Acknowledgments

This book could not have been written without the help of many friends and colleagues. They have encouraged me and offered their help. I hope that all have been listed below.

There are a few chapters that include stories in which I did not participate very much, but these stories played an important role in the Iraq oil sector. For example, the Bayji Refinery story during the surge was very important to the northern oil sector. My participation from Baghdad was minimal, but the story was too important not to include. Input was received from a half-dozen military officers and civilians to document that story. Thanks to input from Dr. Ali Obaidi, Robert Carruthers, Joe DaSilva, Christian Cook, Quentin Coleman, and David Webb, I was able to piece together this important part of the surge.

Another good example was the story about Task Force Shield. My participation with that organization was primarily during the summer of 2003. It was one of our successes of 2003. To get the details of that story, I interviewed retired Colonel Tom O'Donnell in March 2015.

Other interviews included Lieutenant General Jay Garner, Rob McKee, Mike Stinson, Mike Mobbs, Pam Quanrud, Lieutenant Colonel Paul Shelton, Colonel John Agoglia, B. J. Ramos, Dr. Gordon Rudd, Clarke Turner, Robert Maguire, Ambassador Robin Raphel, Lieutenant Colonel Harry Callicotte, US Navy Captain Richard Fritzley, Stephen Lindsey, Kevin Ross, UK Brigadier General Hamish McNinch, UK Major General Richard Cripwell, and Lieutenant Colonel Wayne Dudding. It was a pleasure to meet with each and share our Iraq oil experiences. Some of their stories are included in the book. Ambassador L. Paul Bremer, Dean Popps, and General David Petraeus also provided input.

I was fortunate to find an excellent mentor who had authored more than thirty books, Don Phillips. Don provided invaluable encouragement and assistance to me during the process.

Others helped me with writing the manuscript. They include Mr. Thamir Ghadhban, Ambassador Robin Raphel, Dr. Gordon Rudd, Robert Maguire, Colonel John "J. R." Martin, Pat MacArevey, Tom Mason, Dr. Tom McKenzie, Elizabeth Baldwin, and Dr. Falih Al Jibury. Joe LeMonnier of Mapartist

.com helped with my Iraq oil sector maps. David Larkin provided some partial editing service.

My former oil company colleagues at the Romeo's weekly luncheon group offered encouragement and a valuable source of additional industry discussion.

All statements of opinion or analysis expressed are mine and do not reflect the official positions or views of any US government agency. Nothing in the contents of this book should be construed as asserting or implying US government authentication of information or agency endorsement of my own views.

Abbreviations

ABOT	Al Basra oil terminal
B/D	barrels per day
BUA	battle update assessments
CENTCOM	US Central Command
CERP	Commanders Emergency Response Program
CFLCC	Coalition Forces Land Component Command
CIA	Central Intelligence Agency
CJTF-7	Combined Joint Task Force 7
CMMP	central metering and manifold platform
CPA	Coalition Provisional Authority
DESC	Defense Energy Support Command
DG	director general
DOD	Department of Defense
DOE	Department of Energy
DOS	Department of State
ECON	Embassy Economics Office
EFC	Energy Fusion Cell
EIA	Energy Information Administration
EIPG	Energy Infrastructure Planning Group
ESD	Energy Services Division of J9, US Forces–Iraq
FEED	front end engineering design
FSO	Foreign Service officer
FW	Foster Wheeler
GOI	government of Iraq
GRD	Gulf Region Division of the US Army Corps of Engineers
GT	gas turbine
ICOEE	Iraq Crude Oil Export Expansion
IED	improvised explosive device
IG	inspector general
INC	Iraqi National Congress
INES	Integrated National Energy Strategy
IOC	International Oil Company

IOR	*Iraq Oil Report*
IPSA	Iraq Pipeline through Saudi Arabia
IRMO	Iraq Reconstruction Management Office
IRRF	Iraq Relief and Reconstruction Fund
IT	Iraq Turkey Pipeline
ITAO	Iraq Transition Assistance Office
JCS	Joint Chiefs of Staff
JICA	Japan International Cooperation Agency
JOE	Japan Oil Engineering
KAAOT	Khor Al Amayah oil terminal
KB/D	thousand barrels per day
KBR	Kellogg Brown & Root
KRG	Kurdistan Regional Government
KSA	Kingdom of Saudi Arabia
MB/D	million barrels per day
MoO	Ministry of Oil
MNF–I	Multi National Force–Iraq
MRE	meals ready-to-eat
NGL	natural gas liquids
NSC	National Security Council
OGJ	*Oil & Gas Journal*
OIF	Operation Iraqi Freedom
ORHA	Office for Reconstruction and Humanitarian Assistance
OSD	Office of the Secretary of Defense
OVP	Office of the Vice President
PCO	Project and Contracting Office
PIJV	Parsons Iraq Joint Venture
PMAC	Prime Minister's Advisory Commission
PMO	Project Management Office
PS1	pump station 1
SCIF	Sensitive Compartmented Information Facility
SCOP	State Company for Oil Projects
SIGIR	Special Inspector General for Iraq Reconstruction
SOC	South Oil Company
SOMO	State Oil Marketing Organization
SPM	single-point mooring
TF RIO	Task Force Restore Iraqi Oil
TF Shield	Task Force Shield
TUBA	Tuba pump station
USACE	US Army Corps of Engineers
USAR	US Army Reserves

USF–I	US Forces–Iraq
USG	United States government
VBIED	vehicle-borne improvised explosive device
VLCC	very large crude carrier
WMD	weapons of mass destruction
WQ1	West Qurna 1 oil field
Z1	Zubair 1 pump station
Z2	Zubair 2 pump station, also called IPSA 1

Dramatis Personae

(Sequenced Chronologically)

Washington, DC

Michael Mobbs—Director of EIPG
Pam Quanrud—Oil Contact on the NSC
Clarke Turner—EIPG Team Member from DOE, and ORHA Oil Team
Frosty—EIPG Team Member from CIA, and ORHA Oil Team
Seneca Johnson—EIPG Team Member from State Department
Michael Makovsky—EIPG Team Member from OSD Policy
Lieutenant General (Retired) Jay Garner—Director of ORHA, 2003
Lieutenant General (Retired) Ron Adams—Deputy Director for ORHA, 2003
Major General Carl Strock—US Army Corps of Engineers' Senior Representative on ORHA and CPA
Doug Feith—Under Secretary of Defense for Policy
Paul Wolfowitz—Deputy Secretary of Defense
Dr. Ahmed Chalabi—Chairman of Iraqi National Congress

Iraq

Iraqis

Thaer Ibrahim—Director General of South Refining Company, 2003–2008
Jabbar Al Luaibi—Director General of South Oil Company, 2003–2009
Thamir Abbas Ghadhban—Oil Ministry CEO, 2003–2004, Oil Minister, 2004–2005, Head of Prime Minister's Advisory Commission, 2006–2015
Dr. Radhwan Al Saadi—Director General of Accounting and Economics Directorate, 2003–2007
Dr. Faleh Al Khayat—Director General of Technical Directorate, 2003
Dathar Al Kashab—Director General of Midlands Refining Company, 2003–2010

Mohammed Al Jibouri—Director General of SOMO, 2003, Minister of Trade, 2004–2005
Dr. Ahmed Barifkani—Ministry Downstream Refining Advisor, 2003–2007
Hussain Al Fatal—Director General Distribution, 2003–2004 (killed October 2004)
Abdul Karim Al Luaibi—Oil Minister, 2011–2014 (no relation to Jabbar Al Luaibi)
Riyadh Ghassab—Director General North Refining Company, 2003–2005
Dr. Hussain Al Shahristani—Oil Minister, 2006–2010, Deputy Prime Minister for Energy, 2011–2014
Abdul Jabbar Al Waggaa—Deputy Oil Minister for Upstream, 2005–2007
Ahmed Al Shamma—Deputy Oil Minister for Midstream, 2005–2014
Motasam Hassan—Deputy Oil Minister for Downstream, 2005–present
Dr. Ali Obaidi—Director General North Refining Company, 2007–2010
Dheyaa Jaffar—Director General South Oil Company, 2009–2015
Oday Al Quoraishi—South Oil Company Project Manager, 2006–2011

Americans and Others

Ambassador Lewis Paul Bremer III—Administrator of the Coalition Provisional Authority, 2003–2004
Philip J. Carroll—Former CEO of Shell US and Chairman and CEO of Fluor Corporation, EIPG Industry Consultant, 2002, CPA Director of Oil Operations, 2003
Brigadier General Bob Crear—Commander, Task Force RIO, 2003
Admiral (Retired) David Oliver—CPA Budget Director, 2003
Colonel Tom O'Donnell—Commander of Task Force Shield, 2003–2004
Kevin Ross—CPA Oil Advisor, 2003–2004, Energy Fusion Cell, 2008–2011
John Kjar—Energy Analyst from Australian Government, 2003
Colonel Bob Schroeder—Civil Affairs Officer Assigned to CPA Oil, 2003
Rob McKee—Former Conoco Upstream President, CPA Senior Oil Advisor, 2003–2004
Mike Stinson—Former Conoco Upstream Vice President, CPA Senior Oil Advisor, 2004
Terry Adams—Former BP Executive, CPA Oil Advisor, 2003
Bob Morgan—Former BP Executive, CPA Oil Advisor, 2004 (killed May 25, 2004)
Norm Szydlowski—Former Chevron Executive, CPA Oil Advisor and Embassy Senior Oil Advisor, 2004

General David Petraeus—Commander Multi-National Force–Iraq, 2007–2008
General Ray Odierno—Commander Multi-National Force–Iraq and Commander US Forces–Iraq, 2008–2010
Brigadier General Mike Walsh—Commander Gulf Region Division of the US Army Corps of Engineers, 2006–2007
Dr. Les Dixon—Senior Executive Service GRD Director of Projects, 2006–2007
Captain Richard Fritzley (US Navy Reserve)—GRD Oil Director, 2006–2007
Captain Tom Brovarone (US Navy)—GRD Oil Director, 2007
John Sickman—Embassy IRMO Senior Oil Advisor, 2006–2007
UK Brigadier General Richard Cripwell—Director of Energy Fusion Cell, 2007
UK Brigadier General Carew Wilks—Director of Energy Fusion Cell, 2008
UK Brigadier Hamish McNinch—Director of Energy Fusion Cell, 2008–2009
Admiral Mark Belton—Director of Energy Services Division, 2009–2010
Colonel Greg Zehner—Director of Energy Services Division, 2010–2011

Map of Iraq as of 2015.

CHAPTER 1

American Oil Minister

Who's going to fill the top job?
—Dr. Radhwan Al Saadi

It was April 27, 2003, my third day in Baghdad as part of the initial civilian element that entered Iraq under retired Lieutenant General Jay Garner and the DOD Office for Reconstruction and Humanitarian Assistance (ORHA). Eighteen senior advisors for the various ministries, including me, had arrived in Baghdad from Kuwait on April 24 aboard a C-130 air force transport plane. The other advisors were active and retired ambassadors or other experts in their fields and were assigned as senior advisors to Iraq's ministries. My responsibility was the oil sector, and I was assigned to be the senior oil advisor to General Garner.

Army Civil Affairs Colonel Joe "Bob" Schroeder, USAR, accompanied me in a Humvee to a meeting at the Baghdad Civic Center. The meeting was arranged by my contact at the National Security Council (NSC) of the White House. Colonel Schroeder was assigned to the 352nd Civil Affairs Command, and his mission was to assist the ORHA oil team. Schroeder and his folks were invaluable, especially during my first couple of weeks in Baghdad.

"Gary, do you think these three Iraqis that we're meeting today were some of those that we saw yesterday at the ministry?" Bob asked. Bob and I had met several Iraqis during our first visit to the oil ministry complex the previous day. Bob had taken me to the ministry to initiate contact with Iraqis and meet the US military commander in the area.

My NSC contact phoned me the previous night on my satellite phone. She gave me three Iraqi names for today's meeting. "Bob, Pam Quanrud at the NSC mentioned that these three gentlemen are considered the best oilmen in the ministry. So, I suspect they were among the dozen or so that we met yesterday."

The Baghdad Civic Center was a large complex that became the meeting place for much activity and eventually the home for Iraq's parliament. It was located inside the Green Zone and about a mile from Saddam's Presidential Palace, the current ORHA headquarters. Electricity had not been restored since the beginning of the war a few weeks prior, but the sun was shining and there was plenty of light inside the building. Bob and I quickly moved up the stairs to the second floor toward a small meeting room outside of one of the main auditoriums. As we approached the room, I recognized the three distinguished-looking Iraqi gentleman from the previous day at the oil ministry. A US civilian accompanied them.

Introductions were quickly made and the five of us moved into the small room. The US civilian did not stay. The three Iraqis were Dr. Radhwan Al Saadi (director general of Finance and Economics Directorate), Dr. Faleh Al Khayat (director general of the Technical Directorate), and Mr. Thamir Ghadhban (director general of the Planning Directorate). All were principal directorates within the headquarters of the Ministry of Oil in Baghdad. All three had received their advanced education in the United Kingdom, spoke fluent English, and had at least thirty years of uninterrupted service in the Iraqi oil industry inside of Iraq. Radhwan had struggled to get the other two gentlemen to accompany him to the meeting.

As a twenty-one-year veteran of Mobil Oil and ExxonMobil training, my first item on the agenda was safety. The safety discussion was substantially different from any that I had experienced at ExxonMobil. We discussed the immediate safety of their families and any specific threats to them, especially for meeting with me. The Iraqis were anxious about meeting me privately in the Civic Center because they did not want to be perceived as cooperating with the American occupiers. Many Iraqis considered us occupiers and not liberators, as we were led to believe by our Washington leadership. These gentlemen recognized the importance of making this early private contact. They knew that several of their colleagues in southern Iraq had met with me earlier in April.

The next point on the agenda was the clarification of the US oil sector mission. "Contrary to what you might be reading in the press, we are not here to steal Iraq's oil. Your oil belongs to the Iraqi people," I explained. "The US and my mission is to quickly return Iraq's oil sector to producing at the same level before hostilities started, to repair and restore to prewar levels." And I added with a smile, "So that I can quickly return to my family in the US." This last comment would elude me for several years. April 2003 was my first of seventy-five months in Iraq over the following nine years.

The three Iraqis acknowledged my comments and agreed that we all had a common objective, to get the oil flowing again. "We need to minimize the negative impact to the Iraqi people," said Thamir Ghadhban. "Shortages of cooking and transportation fuels will add to their suffering. Iraqis have already suffered too many shortages in the past under the former regime." Al Saadi and Al Khayat agreed.

"Good, we are in agreement that it is in everyone's interest to get the oil flowing," I said. "Now, what do you recommend I do to help you accomplish the goal?" All three immediately responded that a leadership team must be appointed as soon as possible.

"Until appointments are made, no one knows who is in charge at the ministry," said Dr. Radhwan.

Having sensed that this would be an important agenda item, I reached into my case and pulled out a one-page document that had been provided to me before my departure from DC (see Figure 1.1). It was the organizational structure of the Ministry of Oil, including the names of the individuals in the top twenty-five jobs prior to hostilities. The chart included all the oil ministry directorates in Baghdad as well as the operating companies located throughout the country. It was organized under three deputy ministers: upstream, midstream, and downstream. We focused on all the jobs except for the minister position and the three deputies.

"So, I understand that we need to appoint all of the key leaders designated on this organizational chart. Do you agree?" I asked.

All three Iraqis looked surprised at what was in my hand. "Where did you get that document?" said Dr. Faleh.

"It is not important where I got this, but if I understand you correctly, ORHA needs to appoint Iraqis to these top twenty-five leadership positions within the Ministry of Oil, as soon as possible."

"Yes, that is what we need," said Ghadhban.

"OK, let's get started," I said.

Dr. Faleh proposed three principles before we started evaluating candidates for jobs. "There should be no collective punishments, there should be no retroactive application of new rules, and there should be no guilt by association." He was apparently concerned about possible bias against oil officials who worked under the previous regime and members of the Baath Party. We all agreed, but I interjected that the one exception was involvement with weapons of mass destruction (WMD). I warned that WMD was the one issue where there would be no tolerance. The three Iraqis mentioned that WMD was something that did not involve the oil ministry.

There were no objections, so the five of us spent the next two hours going through each position and discussing the appropriate Iraqi for the job. It was a good discussion, highlighting important skillsets required for specific positions. I was working with professional oil executives, not unlike my colleagues in the global oil industry. Many questions were asked and answered before we settled on a specific appointee for a job. Is he technically competent? What is his educational background? Has he demonstrated leadership skills? How?

Many of the appointees were incumbents and those were easy decisions. A few appointees were designated as temporary or deputy director general

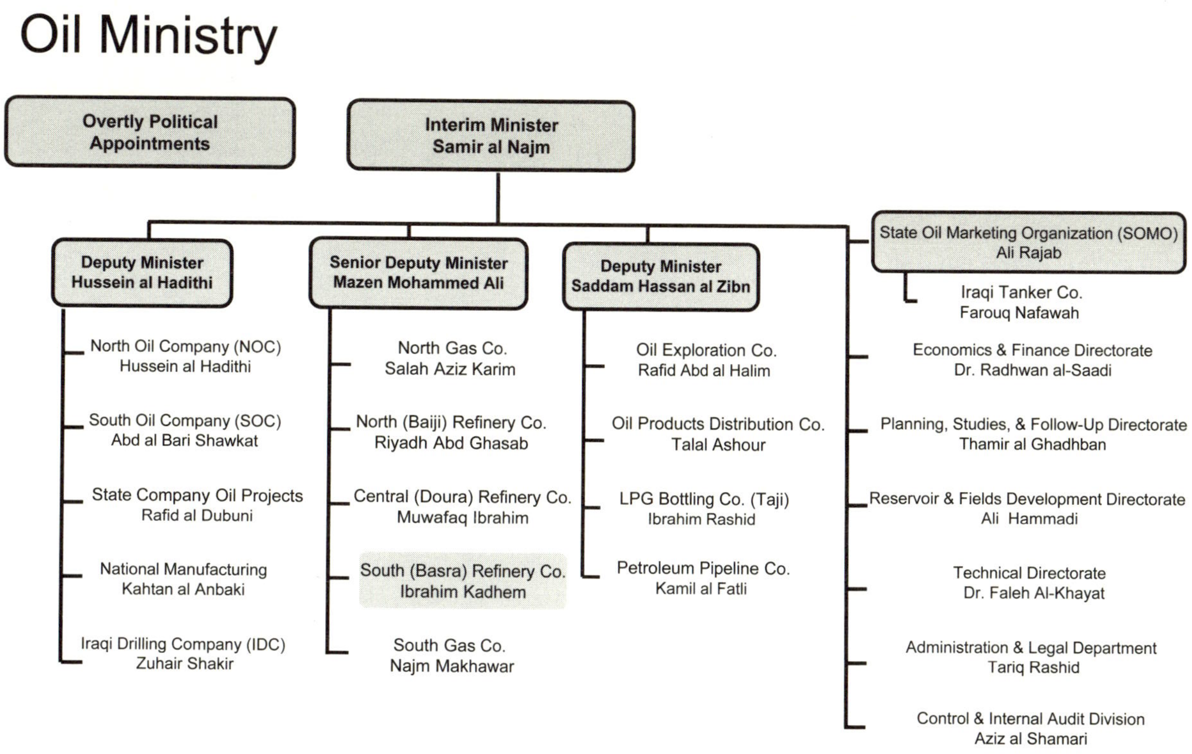

Fig. 1.1 Oil Ministry Organization Chart, March 2003

when it became obvious that no consensus could be reached. Most noteworthy was the appointee for director general of the State Oil Marketing Organization (SOMO). It was one of the smallest groups, but one of the most important. SOMO consisted of about 150 people; most were highly educated and experts in their field. They were responsible for marketing all oil exports from Iraq and had developed an excellent reputation throughout the Middle East oil industry. We decided to temporarily appoint the previous deputy director general.

The Iraqi Tanker Company director was the last appointee.

We agreed to have a meeting on May 3 with the entire leadership team. Formal appointment orders signed by the appropriate ORHA official would be issued for that meeting. The formal meeting would be followed by a public announcement and discussion with the press.

As we stood to leave, Dr. Radhwan turned to me and asked, "Who's going to fill the top job?"

We had discussed all the directors general positions, but we had not mentioned who would be the oil minister, the most important position. The NSC back in DC would make this decision. I would provide input, but the final decision was not mine. So, I decided to get input from the three Iraqis.

"Who do you think should fill that job?" I asked.

Radhwan said he thought it should be an Iraqi who had suffered hardships under the Saddam regime during the previous decades and not an Iraqi expat who had escaped from Iraq years before.

"I will share your recommendation with the people making that decision," I said.

We had all started to walk out of the meeting room when Dr. Radhwan stopped and turned to me, "Gary, until you appoint someone, you are the minister of oil, right?"

He was right. Garner had mentioned that all of his ministerial senior advisors were ministers until an Iraqi was appointed. However, there was something about hearing it come from Dr. Radhwan that made it sound so much more official.

"That's why the appointment will be done quickly. We will have a name by the formal meeting on May 3," I replied.

Dr. Radhwan's comment truly caught my attention. My first thoughts were guarded, almost paranoid. I had just had a telephone conversation with my NSC contact a week prior in Kuwait. Folks in the NSC were irritated about press clippings of US Army general officers quoted as taking credit for early accomplishments in the oil sector. Our orders were to always keep an Iraqi face on anything pertaining to the oil sector. How was I supposed to keep my face out of the press as the American oil minister?

My concerns faded later that evening while cleaning up some of the dirt in my palace bedroom before calling it a day. The room was large enough for about ten army cots and would eventually house five more ORHA oil advisors in another week, but I was the only inhabitant for now.

The windows throughout the palace had been shattered during the "shock and awe" bombing campaign, the massive US air force bombing of Baghdad in the first few days of hostilities. The dust storm that hit most of Iraq the night before had deposited a layer of dirt everywhere. The same storm woke me up at midnight; I was coughing because the dust was so thick that I could not breathe. I grabbed a towel from my duffle bag to filter my air and went back to sleep.

Creature comforts were nonexistent. Meals ready-to-eat (MREs) and bottled water were our nourishment. We had no electricity, no running water, and no indoor bathroom facilities. It did not seem like a palace at the time.

After cleaning things up, I reclined on my army cot and reflected on my new title, "Your Excellency, the Minister of Oil." The title had a nice sound to it—delusions of grandeur, a temporary escape from reality! My new portfolio had proven oil reserves of more than 112 billion barrels; ExxonMobil's oil and gas equivalent reserves were only about twenty billion barrels.

How did I get myself here? What did I do wrong, or right? A kid from East Wheeling, a neighborhood of Wheeling, West Virginia, known for steel mills and coal mines.

I was raised in an Irish-German Catholic family with two brothers and a sister, and religion and sports molded my early life. My father, Henry "Bud" Vogler, was a navy pilot during World War II. He joined the navy a year after graduating from high school and was placed into the navy aviator cadet program. The navy sent him to college for a couple of years and then through flight school. The war ended just as he graduated from flight school, so he left the navy to return to Wheeling and marry his high school sweetheart.

My mother, Margaret Ellen George, was the eldest of six children. Her father fought in World War I with the 42nd Infantry under a very capable military leader, Brigadier General Douglas MacArthur. My grandfather started the Wheeling American Legion Post #1 upon his return from France, one of the oldest in the United States.

Mom and Dad were married in December 1945 and children started arriving a couple of years later. Dad emphasized family values and sports. We played baseball, football, and basketball, depending on the month of the year. Football became the priority sport for us in high school; I was team captain in my senior year.

West Point was my selection for college after receiving a congressional appointment. I jokingly liked to say that I needed a more liberal and less disciplined education for college after twelve years of Catholic school.

The reality was that my motivation for attending West Point was the same as that of General Dwight Eisenhower, to get a free education. My desire to play Division I football also influenced me. Only Division III schools recruited me, so my plan was to "walk on" at West Point and make the football team. During my fifth practice on the West Point plebe team, the coach walked up to me and said, "Vogler, you are too slow. Find another sport to play." I did.

A severe concussion during intramural boxing prevented me from participating in contact sports for a while. So, my attention turned to other activities. I was eventually appointed the editor in chief of our class yearbook, *The Howitzer*. I graduated in the class of 1973, the same class as my more famous classmates General John Abizaid and Ambassador Karl Eikenberry. My appreciation of the West Point experience increased exponentially with age and wisdom. The institution is an asset that has significantly contributed to the success of our great country since 1802.

Going to school with some of the best and brightest young Americans with great leadership potential was a recipe for success. My cadet military unit was company F-2, nicknamed the F-2 Zoo. The Zoo had developed its reputation a few years before my arrival, but we took pride in maintaining our company's traditions. That folklore was steeped with heavy emphasis on the intramural contact sports like football, boxing, wrestling, and lacrosse. It seemed to work well. Just during my four years, company F-2 was the cadet home to such future leaders as General Martin Dempsey (chairman of the Joint Chiefs), Lieutenant General Ben Freakley, Major General Randy Castro, Under Secretary of Defense Mike Dominguez, and Tom Mason, Esq. These five and other Zoo cadets contributed to the success of *The Howitzer* in 1973.

Two weeks after graduation I married my high school sweetheart, Barbara Chapman, and eventually moved to Fort Bragg, North Carolina, for my first assignment with an airborne unit. I enjoyed the camaraderie among the officers in our battalion and especially the pride that went with being a paratrooper.

All lieutenants were expected to attend and graduate from the 82nd Airborne Division jumpmaster course. My turn arrived during my first year. It was a two-week program and the first week consisted of instruction and classroom training. The second week was multiple jumps, with each student placed in the role of jumpmaster at least once. The student would be the jumpmaster while a few classmates would be the jumpers. Graders observed in the back of the aircraft and graded every detailed activity and command given by the student jumpmaster. It took only one small mistake to fail the course. Attention to detail was imperative.

My try as jumpmaster was a success, but during my third jump for a fellow student jumpmaster, disaster almost struck. I nearly crashed and burned. I was the last jumper in a stick of jumpers and was carrying ninety pounds of equipment in front of me below my reserve parachute.

The signal lights next to the open aircraft door changed from red to green. The jumpmaster shouted, "Go, Go, Go," keeping his jumpers moving quickly out the door. His commands of "Go" were barely audible above the loud noise of the aircraft engines. As I approached the door, my equipment caught something either on the jumper in front of me or on the floor of the aircraft, so my jump out the door was not as precise as it should have been; I botched my exit. The wind caught my equipment and started spinning me at a high rate, six or seven revolutions. My parachute had deployed, but it did not feel normal. The riser straps on my parachute were twisted so tight from my spinning that I could not look up to view my chute. I was falling at a higher speed than the other jumpers in my stick.

We were trained to bicycle with our legs when the risers were twisted, to help them untwist. Unfortunately, the ninety pounds of equipment in front of my legs restricted this effort. My fall continued at high speed and I knew that my time was running out. My adrenaline flow was at peak levels. Bicycling eventually worked! My chute fully opened. Looking up, I saw a beautifully inflated parachute.

But the danger was not over. One of the jumpers from an earlier stick was moving under me. All paratroopers realize the hazards of this situation. My parachute collapsed when he moved under me. He took my air. I landed on his inflated parachute. My legs were running as fast as possible, running off the side to get away from him. We were trained to quickly run off an inflated parachute when this happens. Adrenaline had me running faster than ever. I fell quickly toward the earth. My parachute reinflated just a moment before a very hard impact on the ground.

This was the first of many near-fatal events in my adult life. There were half a dozen during my prewar life, but they increased significantly between 2003 and 2011. Either I am extremely lucky or I have a great guardian angel.

My assignments at Fort Bragg taught me several things about myself. The most important was that the army was not for me. The post-Vietnam army of the mid-1970s was not a good army. Drug use and alcoholism among the troops were serious problems. Several soldiers in my first platoon were given the option by law enforcement officials in their hometowns of going to jail or joining the army.

The equipment was old, and senior officers were under constant pressure to make their unit readiness look better than it really was. That said, some of my friendships developed during that first tour have continued for life, like the friendship with one of our country's great orthopedic surgeons, Dr. Tom McKenzie.

My early decision to leave the army enabled me to develop a transition plan. The University of Utah offered an executive MBA program at Pope Air Force Base, adjacent to Fort Bragg. It was one of the first of its kind and many junior military officers took advantage of the program. Robert McDonald, former secretary of veterans affairs, former chairman of Proctor & Gamble, and West Point class of 1975, took advantage of the same program during his first assignment at Fort Bragg.

With my West Point engineering degree and MBA, a civilian career in late 1977 was my preference. Hold on. The army had other ideas, and one year remained on my five-year West Point commitment. The army personnel office advised me that an overseas tour was necessary. My options were a one-year, unaccompanied (without wife) tour to Korea or a three-year, accompanied tour to Germany. Guten Tag, Frau Vogler, we are going to Germany!

The three years in Germany passed quickly. A recruiting agency prepared me for transitioning to a civilian career. Our daughter, Lindsay, was born at the Landstuhl military hospital in April 1980, and we all returned to the United States in December 1980.

After just a few corporate interviews and offers, Mobil Oil was my choice. They sent me to their petrochemical facility in Beaumont, Texas. The facility was new and Mobil was planning major expansions over the following five years. I stayed for more than six years. During those years, we saw the addition of our son, Chase. Jobs in the production, engineering, and contracts departments provided good junior management experience.

Mobil sent the family and me to Saudi Arabia in 1987 for an opportunity at the new Yanbu joint venture export fuels refinery. This assignment helped prepare me for my future Iraq work. The exposure to Middle East culture, refinery management, and oil pricing gave me skillsets that eventually served me well in Iraq.

The first Gulf War occurred during my tour in Yanbu. Mobil immediately evacuated all the families; the employees stayed on the job. Our refinery became the primary supplier of military fuel to the US military for the war. We were located in theater but far from Kuwait on the Red Sea coast. My army reserve unit at the time of the first Gulf War was the EUCOM HQ in Stuttgart, Germany. My mobilization to active duty in Stuttgart to support the war effort was canceled when EUCOM recognized the role of our refinery as a primary supplier of fuel for the war.

The four years in Yanbu were enough to get me assigned to the Mobil Oil headquarters in Fairfax, Virginia, in September 1991. Several staff positions followed for me. One of the highlights of my Fairfax career was being part of a great American corporate turnaround. Mobil's US Marketing and Refining Division undertook several cost-cutting and reorganization initiatives in the early 1990s to move the financial performance from worst to first among its US competition. That competition included Exxon, Amoco, Shell, Texaco, Chevron, and BP. Mobil maintained that leadership for five years until the merger with Exxon.

David Norton, creator of the "Balanced Scorecard," has featured Mobil Oil's corporate success story in several of his books. A professional highlight for me was being placed on the implementation team with Norton's Renaissance Solutions consulting group.

One of my jobs during that time was to advise our division president daily on key performance indicators and implement the scorecard strategy. The lesson I relearned during those years was the same lesson drilled into me at West Point: it's all about leadership.

My post-merger position with ExxonMobil was a good job. I was the planning manager for a Global Procurement organization with more than three thousand employees. The US Government Accounting Office (GAO) considered our company's Procurement organization as one of the three best in the country. I participated in a benchmarking exercise with the GAO in 2000. The GAO was charged with recommending changes to government procurement based on industry best practices.

My twenty-one-year career with Mobil Oil and ExxonMobil ended in February 2002 when my request to receive the executive financial separation package was approved. This "golden parachute" provided the financial incentive to exit. It was available to Mobil Oil executives departing within two years of the merger.

The ExxonMobil culture was not for me. Decentralization and empowerment had been keys to success under Mobil Oil, but control and centralization best described the new culture under ExxonMobil. I had a great boss, Dr. Jean Baderschneider, vice president of Global Procurement. Jean had the sharpest mind of all of my previous supervisors. This made my decision to leave very difficult, but I caught myself counting the months until retirement. A successful fight with cancer in 1996–1997 reminded me of my mortality; life was too short to be counting down months until retirement. My last six months were spent in Indonesia to fix some company problems. An Indonesian insurgency had attacked our pipelines and shut down operations. In return for taking this six-month assignment, my reward was the financial separation package. My experience in Indonesia with fixing pipelines after insurgent attacks would eventually serve me well in Iraq.

My army reserve assignment in the late 1990s and in 2002 was as a military academy liaison officer working with West Point admissions and high school students in northern Virginia. This work was very satisfying. Five seniors at Centreville High School landed appointments to West Point in 1997. Similar to many of my northern Virginia appointees, those five individuals graduated from West Point, served tours in war zones, and continued their careers with the army. As of 2015, all five had achieved the rank of major and continued their career in the military.

My West Point work also helped me network with fellow graduates in the DC area. It led to an additional army reserve assignment with the headquarters at the Army Corps of Engineers. My supervisors in that unit attended a meeting at the Pentagon in early fall 2002 that discussed an Iraq oil planning effort. They forwarded my resume and that initiated my Iraq appointment.

On Wednesday, October 23, 2002, I went for an interview at the Pentagon after being told by my army reserve unit that I was needed there by the Office of the Secretary of Defense (OSD). My only question was, "What is OSD?" The Office of the Under Secretary of Defense for Policy (USD-P) was going to be my new workplace for a while.

The Pentagon was an impressive structure from the outside, but walking the corridors on the inside was inspiring. Pictures of famous military leaders and military hardware lined the walls. My eyes were wide open as I was quickly escorted through the halls and into an interview.

My interview that day in October 2002 was with Michael H. Mobbs. Mike had just been appointed to lead an interagency planning cell, the Energy Infrastructure Planning Group (EIPG). Mobbs was a lawyer and protégé of Doug Feith, the undersecretary of defense for policy. His law degree was from the University of Chicago Law School. He was brought into the Pentagon with the Bush Administration and had been working on several issues, including the prison issue at Guantanamo Bay.

Mike told me to take a seat as he sat down and said, "Gary, let me get to the point. I have reviewed your civilian resume and I think you can help us. We have been tasked by Secretary Rumsfeld to put together an interagency planning cell with participation from State, CIA, and DOE. Do you have a current security clearance?"

My army reserve security clearance was up-to-date at the SECRET level. I mentioned this, but I suspect that he already knew the answer since he had been in contact with the Corps of Engineers.

"What are the terms of reference for this planning cell?" I asked.

Mike pulled out a document with the bright red secret cover sheet. He turned the document for me to read. It was a one-page letter signed by Secretary Rumsfeld outlining our task.

Mike gave me a couple of minutes to digest the contents of Rumsfeld's letter. He told me that there were no plans to activate me within the army reserves for the work. He was bringing me in on a contract for a period of no more than sixty days. He said that no one currently assigned to the team had oil industry experience like mine. The DOE person was scheduled to arrive from Wyoming in a few days and he operated oil wells for DOE out west, but he had never worked for an oil company. Mike then asked me if I was interested.

I looked down at the document one more time and realized the importance of what he was asking me to do. "Mike, when do you want me to start?"

That was the beginning of five months of prewar planning followed by seventy-five months in Kuwait and Iraq, until my final departure from Baghdad at the end of September 2011. I had a seventeen-month tour (March 2003 through July 2004) and a fifty-eight month tour (December 2006 through September 2011) under the Department of Defense. I returned to Iraq in 2015 for an assessment of the oil sector under a USAID contract. All my tours were adventures; some had more adrenaline rushes in a month than most Americans experience in a lifetime.

As a civilian, I had no political or agency loyalty issues. My politics were centrist, with no chosen political party affiliation. My parents were Kennedy Democrats, but my grandparents had been staunch Republicans. As a successful ExxonMobil employee for twenty-one years, an army reserve officer (rank of lieutenant colonel) for more than twenty years, and a West Point graduate, I was well qualified for the demanding assignment. Some of my observations in the oil sector prompted questions at the time, and some of those questions remain unanswered or only partially answered today.

What role did oil play in the decision to go to war? What is the Haifa pipeline and why did the press question the oil minister so much about this pipeline? What were some of the oil agendas and what impact did they have on the mission? Why did our bombing targets avoid all oil infrastructures except for an oil pump station in Anbar Province? Why did the Iraqis attack oil pipelines? Was the prewar oil planning effort effective? How did the ministry headquarters get spared from the looting and burning? Why was fuel supply the top priority in 2003? What role did oil play in the surge of 2007? What is the untold story of Iraq's auction of their large oil fields in 2009?

The answers to these questions will be discussed in the following chapters. The search for answers to these questions prompted troubling thoughts for me. We lost 4,489 American military personnel between the start of 2003 and the end of 2011. Many faces of those young men and women that made the ultimate sacrifice were etched in my brain. More than thirty thousand American servicemen and servicewomen received severe, life-changing injuries. The Treasury estimated that the total cost of the Iraq War would pass $2 trillion when veterans' costs are paid.[1] This was a huge cost to our country for what currently appears to be of little to no benefit.

There are a couple of messages throughout this book. First, partnering and teamwork between Iraqis and the coalition enabled us to overcome huge obstacles and achieve successes that kept the oil flowing. Most of those obstacles were self-imposed because of American arrogance. Despite a lack of military resources on the ground, Washington political agendas, CPA's payroll and budgeting policies, government procurement delays, a Syria agenda, and other problems, the oil continued to flow. Even the Iraqi obstacles of the insurgency, criminal activity, and political decisions were not enough to stop the many successes in the Iraqi oil sector. This book covers most of those successes in detail.

Second, I am now convinced that oil did play a role in the decision to go to war. I would not have made that statement prior to 2014. Learning several facts changed my mind. It is my opinion that a dream to provide Iraqi oil to Israel played a role in the decision to go to war. How much of a role is not known, but shipping oil worth $30 to $50 billion through Israel every year would have enriched many and solved Israel's serious energy problems. Several chapters discuss this oil agenda.

CHAPTER 2

Contingency Planning for the Oil Sector

The road to hell is paved with good intentions.
—Popular Proverb

In preparing for battle I have always found that plans are useless, but planning is indispensible.
—General Dwight D. Eisenhower

The Bush Administration was criticized for a lack of prewar planning for Iraq after the hostilities. The oil planners took exception when it came to the Iraqi oil sector. The planning for the oil sector was good. About 95 percent of Iraq's foreign revenues came from oil sales. If the oil did not flow, then Iraq did not have an economy. It would have been irresponsible not to plan for reconstructing the oil sector.

What were the legal obligations of the United States in regard to Iraq's natural resources, especially oil, after becoming an occupying power? What must the United States do under international law as an occupying power? What may the United States do? The Bush Administration started thinking about these and other similar questions in 2002, as the drums for war started beating louder.[1]

An interagency planning cell, CENTCOM staff, and eventually the US Army Corps of Engineers developed a contingency plan in late 2002 and early 2003. In retrospect, more than ten years later, it was a good plan.

My first full workday with the interagency planning group was Friday, October 25, 2002. Mike Mobbs mentioned that our name was the Energy Infrastructure Planning Group (EIPG). The word "oil" could not be used in the name; the word was deemed too sensitive to be placed on the outside door of the planning Sensitive Compartmented Information Facility (SCIF) on the ground floor of the "A" ring in the Pentagon. That was the first indication that there was a lot of sensitivity to any optics (public sentiment) concerning the Iraqi oil sector. The Bush Administration consistently downplayed an oil motive for attacking Iraq. Also, the work had to be kept very confidential so as not to undermine any diplomatic efforts being undertaken.

EIPG was organized as an interagency planning cell reporting to the National Security Council (NSC). Stephen Hadley, Frank Miller, and Pam Quanrud were our contacts at the NSC. Pam Quanrud was our primary contact. The State Department, Department of Energy, CIA, and Office of the Under Secretary of Defense for Policy provided members for this cell.

The formal reporting was to the NSC; however, reporting to Doug Feith's office took place before sharing anything with the NSC. Specifically, Abrum (Abe) N. Shulsky and Doug Feith reviewed our work at the Pentagon. I. Lewis (Scooter) Libby was the chief of staff for Vice President Cheney. Libby reviewed our work on occasion. Their comments were incorporated into any documents before they were shared with the NSC. Agency representatives on the team shared them with their specific agencies.

EIPG consisted of eight full-time participants, one part-time participant, and several people that were brought in at various times for specific reasons.

The following were the eight full-time participants:

- Michael Mobbs—group lead; reported to Doug Feith, undersecretary for policy
- Clarke Turner—Department of Energy
- Seneca Johnson—State Department
- Frosty—CIA
- Michael Makovsky—Office of Under Secretary of Defense for Policy
- Barbara Glotfelty—DOD procurement contracting officer
- Matt Amitrano—DOD contractor, administrative assistant
- Gary Vogler—DOD independent contractor

Michael Mobbs graduated from Yale University in the early 1970s and later received his law degree from the University of Chicago Law School. From 1982 until 1985, Mike served as the secretary of defense representative to the strategic arms reduction talks for Assistant Secretary Richard Perle. It was during this time that Mobbs first worked with Doug Feith.

In December 1985, President Reagan appointed Mobbs as assistant director (strategic programs) of the US Arms Control and Disarmament Agency. In 1987, Mobbs left the Reagan Administration and joined the law firm of Feith and Zell, located in Washington, DC. Mobbs spent several years in Russia in the late 1990s working in the private sector, returning to DC before the Bush Administration.[2]

Doug Feith was confirmed by Congress and appointed as undersecretary of defense for policy in the spring of 2001. Feith then approached Mobbs and others about joining his team at the Pentagon. The original plan was to bring in four experienced people with whom Feith had gained confidence: Abe Shulsky, Sven Kramer, Michael Pillsbury, and Mike Mobbs. This group would act as a "think tank" for Feith and be assigned strategic projects by the undersecretary. Then 9/11 hit. The think tank plan was overcome by events

(OBE is the military term). All four were quickly brought on board. They received security clearances and were placed into positions. Mobbs started in November 2001.[3]

Mike Mobbs came from a family where all the male members had served in the military. He was the first who did not serve in uniform. His father fought in World War II along with his uncles. His son, also named Mike, was a freshman at Tulane University in the spring of 2001 but wanted something different in his life. He wanted to be a Navy SEAL or an Army Ranger. Mike Jr. quickly joined the Army Rangers after the 9/11 attacks. The army sent him to hot spots in the war on terror during 2002. Mobbs knew that his son would be one of the first soldiers entering Iraq, if a war started. Mobbs had skin in the game.

Mobbs denied that his son's military service influenced his thinking during EIPG, but I firmly believed that it had to have some impact. His objectivity, criticism, and voice of reason among a group of political appointees rushing to war were commendable. He pushed back against those that did not have any skin in the game. Mobbs mentioned to me that he was never personally convinced that the war was necessary or that war was the right decision. No one ever asked his opinion and it was not his job to give such an opinion.[4]

Clarke Turner was the Department of Energy representative on the team. He was the director of the Rocky Mountain Oil Testing Center (RMOTC) facility in Casper, Wyoming. Clarke was originally from New York, graduated from the Colorado School of Mines in the early 1980s, and joined the US Army for several years as a petroleum management officer. He had excellent petroleum supply experience in Korea and during Desert Storm in 1991.

After leaving the army, Clarke joined the Department of Energy and eventually became a director in Wyoming. The RMOTC had more than one thousand oil-producing wells in 2002. The EIPG team valued Turner's experience in the military as well as in the Department of Energy. We worked closely together until his return to RMOTC in July 2003.

One distinguishing physical characteristic of Clarke was a limp in his gait. Clarke belonged to a small and very blessed group of individuals: survivors of grizzly bear attacks. An attack left one leg mangled, saved only after several months in a hospital and multiple surgeries. I felt a common "survivor" attribute with Clarke that contributed to my confidence in him.

Seneca Johnson was the State Department representative on the team. She was a Foreign Service officer (FSO) with several years' experience in the Bureau of Near East Affairs at the State Department. Seneca did not go to Iraq initially. She was pregnant with her son. However, she later took an important assignment in Baghdad during 2011 as Ambassador James Jeffrey's chief of staff.

Seneca had the most US government Middle East experience of anyone on our team. I always sought out her opinion and placed high value on it. When I asked her opinion of what I thought was a rush-to-war mentality, she made a prophetic comment. She quoted the popular proverb that the road to hell is paved with good intentions. That comment came to my mind many times over the next decade, especially in 2006 and 2007, when Iraq was imploding.

The representative from the CIA was a person known as Frosty. (His complete name is intentionally not being mentioned.) Frosty brought an enormous amount of Iraq oil sector intelligence to the team. His personality and chemistry matched very well with Clarke and me. His value to our team cannot be overstated.

Barbara Glotfelty was a DOD professional with more than twenty years in the civil service. She was our contracting officer and assigned to our team by the DOD acquisitions group.

Michael Makovsky joined the office of the undersecretary of defense for policy in the summer of 2002. He reported to Abe Shulsky. In his late thirties, Mike was a young person whose background in oil and gas was minimal. He was always open to discuss his religion (Orthodox Judaism) and his youth in St. Louis, especially Cardinals baseball. We had many interesting and informative discussions on both subjects. However, Makovsky was secretive about his adult life and spoke only of his recent PhD schooling.

Clarke Turner and I approached Makovsky early about his oil and gas background, so that we could better understand his role on the team. Makovsky's response was that he was assigned to the team because he was Doug Feith's brother-in-law. That was not the case. We took that response to mean his background was none of our concern. We eventually learned that his limited oil and gas experience was working for a small oil-trading firm somewhere in the northeast for a short period.

Mike Mobbs said that when he was appointed as the head of EIPG, he looked for anyone with any experience in oil and gas in OSD Policy. He was told that Makovsky had contacts in the oil industry so Mobbs agreed to take him. Mobbs mentioned during my interview with him in 2015 that the American Israel Public Affairs Committee (AIPAC) helped place Makovsky in Shulsky's group during the summer of 2002.[5] Although Makovsky's experience was limited, Mike Mobbs and the team used his strengths in a positive way to contribute to our success.

Makovsky was smart and had good writing skills. We perceived that he had some type of personal relationship with Feith and Wolfowitz. He encouraged this perception by frequently mentioning their names. Makovsky told us that Feith had helped him with his PhD and his publication on Winston

Churchill. Clarke and I were skeptical of his proclaimed close relationship to Feith, but we could not identify any other sponsor.

Makovsky seemed to have good information to support his inner-circle relationships. Here are some examples:

- During an early discussion of possible Iraqis to lead the oil ministry, Makovsky removed Dr. Ahmed Chalabi from our short list for consideration. There were bigger plans for Chalabi. Makovsky said that Chalabi was supposed to be the first post-Saddam prime minister. This was the first of many experiences I had that pointed to Chalabi being the anointed post-Saddam leader of Iraq. Chalabi was the controversial leader of the Iraqi National Congress (INC), which provided much of the false intelligence to the United States about the existence of Iraqi weapons of mass destruction.[6] Wolfowitz and Feith promoted Chalabi, but they never convinced Rumsfeld to publicly support their selection. Secretary Rumsfeld would later deny in his book that he had selected any specific Iraqi to lead Iraq.[7]
- After a senior Foreign Service officer, Tom Warrick, was invited to EIPG for an enlightening discussion on the State Department's oil work in Iraq, Makovsky let us know that Tom was despised by Wolfowitz and Feith and should not be invited back to the Pentagon. Tom had led an independent State Department "Future of Iraq" project: networking with Iraqi expatriates other than Chalabi's INC group to study various sectors of Iraq, including oil.
- After Makovsky returned from being a backbencher at a high-level OSD meeting between civilians and senior general officers, he mentioned that the generals did not want to go to war. This comment took me by surprise, as if going to war were something that military people should enjoy. His comment immediately prompted a speech from me on the seriousness of going to war, but it also let me know that he was probably just echoing the senior leadership tone of Feith and Wolfowitz.
- Battles between DOD and DOS were not only at high levels but also at lower levels when the opportunity became available. After a meeting with State Department Foreign Service officers when Michael Rubin (an OSD colleague of Makovsky) abrasively dominated the meeting, Makovsky let us know that Rubin was the junkyard attack dog.
- There was widespread distribution of an unclassified report from the CIA on possible scenarios for Iraqi society after the fall of Saddam. It depicted a high probability of tribalism, sectarianism, and violence.

> Makovsky let us know that any reports like this should never officially be shared with Feith and Wolfowitz.

Makovsky had problems getting a security clearance. Walking into our SCIF one day, he was visibly angry about something. I asked what was bothering him. He said that he had his third security interview for his top-secret security clearance and the Pentagon security team wanted more information from him. "Three interviews," I said. "Wow, you are the first person I have ever met that required more than one interview. You must have an interesting past." He responded that Pentagon security was interested in his friends in Israel and that he was going to Feith about this problem. His top-secret security clearance was approved about a week later.

The mystery of Mike Makovsky's background was partially solved in 2005. During the summer of 2005, a person who had worked with me in Iraq and was familiar with the problems with Makovsky during the summer of 2003 sent me some Internet information. The Internet blog referenced a *St. Louis Post* interview written on Makovsky in 1989. The *Post* article discussed Mike's plans to "launch a career in Israel's foreign service." The interview also mentioned that "he'll have to serve a stretch in the [Israeli] army before entering the country's foreign service."[8]

My contacts within our government's intelligence agency told me that Israel's "foreign service" could mean the Israeli intelligence, Mossad. The tone of the article highlights a twenty-five-year-old American male who wanted to commit his future to the country of Israel. The article quotes him: "But I have strong feelings about helping to build a Jewish state." As a military man myself, I personally respect anyone who swears allegiance and their life to their country. Upon seeing this article and remembering the problems that Phil Carroll and I had with Mike in the summer of 2003, I immediately phoned Phil to see if he was aware of Makovsky's background. It was during this conversation with Phil that I learned that Mike's top-secret security clearance was awarded to him because Doug Feith "trumped" the questions and concerns of the Pentagon security investigators. It was the first time I realized that such actions were even legal.

The next time I saw Makovsky after reading the *Post* article was in December 2005 at a meeting at the State Department. When I asked about his service in the Israel Defense Forces (IDF), his response was that he refused to confirm or deny such accusations. It sounded as if he had lawyered up. There was an AIPAC espionage scandal that was occurring at the time in Feith's office at the Pentagon. Lawrence Franklin was sentenced in January 2006 to twelve years for espionage-related charges. Franklin worked in Feith's Northern Gulf directorate.[9] This may have been why Makovsky sounded so guarded.

Unlike the other EIPG team members, Makovsky refused to go to Iraq, remaining at the Pentagon in a junior staff position under Doug Feith. He had a different agenda, and his agenda proved to be a distraction to our mission. Several incidents corroborate this impression and they will be discussed later in the book, such as how Makovsky verbally attacked the head of the Iraqi oil ministry, Thamir Ghadhban, and threatened Major General David Petraeus in the summer of 2003 regarding Syrian oil products. He was also quoted in a book by Steve Coll as saying that his high-priority project in the summer of 2003, the Haifa pipeline, was in response to an Israeli request.[10]

Clarke Turner said that Makovsky was different from other young staffers brought in by political appointees he had worked with in the DOE. Mike tried to bully people, but there was nothing behind his bravado. "His interest was Syria and especially the pipeline going through Syria from Iraq," Clarke said.[11]

We had one EIPG participant who worked part-time. He maintained a desk in our SCIF and his name was Mark Feldman. He had worked at the State Department in the 1970s and joined the law firm of Feith and Zell in the late 1980s. He participated in team meetings when he was present. It was not clear if this work was done pro bono or if he was a paid consultant reporting directly to Doug Feith.

There was a group of individuals periodically brought in by Mike Mobbs as subject-matter experts. They were oil industry experts who had retired from the industry, and most had participated in the reconstruction of the Kuwait oil sector after the first Gulf War. They were brought in to review the Contingency Support Plan (CSP) for Iraqi oil reconstruction. A couple of these experts were retired from oil service companies and had actually worked inside of Iraq during the 1980s. They were American citizens with security clearances. Mike Mobbs was able to rush their security clearances through the Pentagon bureaucracy in just a few weeks so that they could review the draft iterations of the CSP. To save on time and travel expenses, Mobbs coordinated with government installations close to where these experts lived so that the draft CSP's could be reviewed at a US government SCIF in their area. These gentlemen were not compensated for their time, but they were reimbursed for their travel expenses.

Others were brought into our office as guests of Makovsky to discuss the Iraq oil sector and to offer suggestions for marketing Iraq's oil. These individuals were not retired. They were active employees of private sector companies, small companies from the northeast. Only one active international oil company employee entered our SCIF. He was working for Hess Oil at the time and was a personal friend of Makovsky. To my knowledge, none of these individuals possessed a security clearance and none was reimbursed for expenses by the government.

The workload for EIPG was divided into two major categories: policy decisions required by the president's cabinet, and the contingency plan for Iraq's oil sector.

In regard to the first category, policy decisions by the cabinet, there were several policy decisions that needed discussion and agreement at the highest level of the US government before going into Iraq. Some of the issues included the following:

- What will be our guiding principles and legal considerations?
- What was the US mission related to the oil sector?
- How should the oil sector be organized and managed?
- What were the production targets?
- How would oil be sold and revenues captured for reconstruction after removing Saddam?
- What should be the relationship during the interim government with Iraq's neighbors, including Jordan, Saudi Arabia, Turkey, Kuwait, Syria, and Iran?
- What should be done about UN sanctions after the fall of Saddam?

Our task was to suggest our policy decisions to the agencies and through the NSC for initial presentation to the Deputies Committee. The Deputies were the "second-in-charge" of each major agency and included Paul Wolfowitz from DOD, Richard Armitage from DOS, and others from the CIA, Treasury, Energy, and the like. Upon approval by the Deputies, policy recommendations were presented to the president's cabinet for final approval.

Mike Mobbs decided to take each policy decision and assign it to one of the EIPG team members. Some team members were assigned more than one. Each of us took our policy decision and followed a PowerPoint format that we had agreed on in order to present the policy to the decision-makers. This format was anywhere from six to ten PowerPoint slides that included a detailed discussion, the pros and cons of options, recommendations, conclusions, and the follow-up required. The original draft was first staffed within EIPG followed by reviews with Doug Feith. After comments were incorporated, a final product was sent to the NSC so that a review by the Deputies could be scheduled.

This work started during the first week in November and was completed before the end of 2002. As a person from the industry, I was impressed with the way in which all the stakeholders weighed in with their opinions. Similar to in the industry, some opinions and comments received much more weight than others.

One policy was not handled in the same way as the others. It was a policy that involved the relationship with Syria and other border countries after the fall of Saddam. When Mike Mobbs assigned that specific policy, my suspicions were aroused. Mike Makovsky insisted that he take the lead on that particular policy. His aggressiveness during this part of the meeting caused everyone to stop, a pregnant pause. The subject was more appropriate for State, but no one objected, and so Makovsky took the lead on that policy. The details are noteworthy.

I was in Mobbs's office discussing another subject when Makovsky walked in and handed Mobbs his first draft. Makovsky left the office and Mobbs thumbed through the report quickly. Mobbs looked at me and asked me if we were going to war against Syria or Iraq. The look on my face must have been enough. Mobbs handed me the multipage PowerPoint draft that Makovsky had just handed him. I reviewed it quickly and understood Mobbs's comment. It was extremely hawkish toward Syria, and in no uncertain terms Syria was to be treated as belligerent in post-Saddam Iraq.

Makovsky's draft received significant changes before leaving EIPG and was drastically changed after incorporating comments from the DOS. The final presentation approved by the president's cabinet was totally different from Makovsky's initial draft. However, the Syria policy espoused by the OSD civilians a few months later was much closer to Makovsky's initial draft than to the cabinet's final approved version. For example, Paul Wolfowitz verbally reversed written instructions and cabinet-approved policy during a meeting with General Franks in March 2003, before hostilities started. This occurred after Secretary Powell had made his convincing argument to the United Nations that Iraq had WMD. The change made by Wolfowitz led to the destruction of the large pump station that pumped Iraqi oil exports through Syria. This is discussed in more detail later in the book.

The second category of our workload was the contingency plan for Iraq's oil sector. Clarke Turner, Frosty, and I spent much of our time on this portion of the EIPG mission. It took a few weeks to get this portion moving because of significant political sensitivities about using Kellogg Brown & Root (KBR) for our work.

The CEO of Halliburton/KBR had been Vice President Cheney before the Bush Administration. KBR had competed and won the multibillion-dollar Logistics Civil Augmentation Program (LOGCAP) III contract in 2001 to support the US Army in the war on terror in Afghanistan, Iraq, Kuwait, Djibouti, and Georgia. The scope of this contract covered engineering and construction, as well as many other services. It was truly the "catch-all" logistics contract for the army and could especially be used when "time was of the essence" and the army did not have three to six months for a competitive bid process.

General Tommy Franks was pushing his CENTCOM staff hard. Franks wanted to be ready for war in mid-February. The actual start date ended up being March 20, 2003. The decision was made in November to issue a task order under the LOGCAP contract with Halliburton/KBR to develop a contingency plan for reconstructing the Iraqi oil sector. The decision to use KBR was the best choice that EIPG could make.

To their credit, KBR quickly mobilized resources in their Houston office to meet the requirements of their task order. Leading the group for KBR were two former US Army officers. Colonel (retired) Ray Rodon was the project manager developing the plan. He eventually went to Kuwait with KBR a few months later for the execution of the plan. The lead writer of the plan was Betty J. (B. J.) Ramos, an army reserve officer. KBR also sent Ramos to Kuwait a few months later. Subject-matter experts from all over Halliburton and KBR volunteered to be part of the team. All of these volunteers had prior military experience, some for just a couple of years. All KBR employees assigned to the project had US government security clearances.[12]

The location of the plan development was in KBR's Houston offices. KBR had to maintain a US-government-certified security office for their classified work under the LOGCAP contract. All planning material was controlled and secured as if the plans were completed at the Pentagon. All employees knew that discussion of the classified plan outside of the secure office was a violation of federal law. The security of the planning document was taken very seriously by all involved.

EIPG's contact on the CENTCOM staff for oil sector planning was Marine Lieutenant Colonel Paul A. Shelton, J5 Plans section. Paul visited our Pentagon SCIF in November along with Cliff Fowler (CENTCOM J2) for initial introductions. Almost all of our work with CENTCOM was with three staff sections: J2 (intelligence), J3 (operations), and J5 (plans). Paul was a graduate of the Naval Academy and chose a commission in the Marine Corps upon graduation. Cliff was a career civilian intelligence expert who had worked at the CENTCOM HQ in Florida for many years.

Since KBR had about two-dozen subject-matter experts in Houston, it was decided that EIPG and Lieutenant Colonel Shelton should travel to Houston to meet with the planners. We did so immediately following the award of the task order.

One of the more interesting stories was an initial meeting around a large table with Clarke Turner, Frosty, Lieutenant Colonel Shelton, several subject-matter experts from KBR, and me. The engineers/subject-matter experts were asking a lot of questions—making a lot of demands on the military as to what they considered important and critical.

Paul Shelton had been schooled in military strategic planning. He decided to share his military cost-benefit analysis training with the engineers who were requesting so much information.

"Okay, how many eighteen-year-old lives is it worth for us to have this infrastructure? That's what we are talking about, killing eighteen-year-old Iraqis and losing eighteen-year-old Americans to secure these components. You need to be very sober about this. If it's not critical, don't tell me it's critical, because people may die to get it." There was silence in the room when Paul finished his comment. He got their attention. Such was the very serious reality of planning for war!

KBR completed their task and delivered a draft plan to the EIPG offices at the Pentagon a week later. In anticipation of the plan, EIPG scheduled meetings for several industry experts to review the plan. A couple were retired CEOs. The rest were retired executives with oil companies and oil service companies. Almost all of these executives had experience in Kuwait after the first Gulf War.

The comments, suggestions, and recommendations made about the KBR original draft plan were excellent. One of the retired CEOs who spent an enormous amount of time and developed several pages of suggestions and recommendations was Phil Carroll. Phil was the former CEO of Shell US, as well as the former CEO and chairman of Fluor Corporation. At sixty-five years old, he was an elder statesman in the US oil and gas industry, and he was friends with President George H. W. and Barbara Bush.

Carroll had spent a few years in the military back in the 1950s. He had taken army ROTC while a student at Tulane University in New Orleans. This experience added to Phil's value as a key contributor to our contingency plan. His comments, questions, and discussion during our meetings impressed everyone on the team. Knowing that we needed a senior-level person from the industry in Iraq during the execution phase, his name was placed at the top of a short list. Mike Mobbs asked me if I thought Phil would be willing to join us in Iraq. I responded, "I don't know how any patriotic American could turn down a personal request from the secretary of defense when our country was going to war." Phil received such a telephone call from Secretary Rumsfeld in March 2003.

After all comments and recommendations were incorporated, KBR submitted their final plan. The plan was in three-ring binders and on CDs. There were fourteen appendices that were well over two thousand pages long. It was an excellent document for a war plan. The format was in the standard military operation order format. Even if we had another year to plan for the oil sector, a significantly better plan for oil would not have been achieved.

Much was written on oil well firefighting. Our most likely scenario included 250 well fires in the south and 250 in the north that would have been our major focus for the first six to nine months. Fortunately, that did not happen.

The plan was declassified in 2008 and can be found on a Pentagon FOIA (Freedom of Information Act) website—Contingency Support Plan (CSP), declassified May 8, 2008.

One of the final actions of EIPG was to write the formal action memo from Doug Feith to Secretary Rumsfeld on January 16, 2003, declassified in 2015. This memo assigned the Department of the Army as the executive agent for the oil program. A multipage PowerPoint presentation was attached to the memo detailing policy issues. Besides appointing the army, the significant issues covered in the presentation were the following:[13]

- Petroleum Proceeds—The annual Iraqi oil revenues were forecasted to be $16 to $25 billion. The uses of the revenues included the occupation at $48 billion, humanitarian aid of $16 billion, and other demands that resulted in a deficit of $100 to $140 billion per year.[14] Paul Wolfowitz testified before Congress in March 2003 that "we're really dealing with a country that could finance its own reconstruction."[15] His testimony to Congress was not consistent with the work we accomplished in EIPG.
- Candidates Proposed to Lead Oil Sector—Several high-profile Americans were proposed to be chairman of the Iraqi oil sector. The list included James Schlesinger, James Baker, Frank Carlucci, and General Norman Schwarzkopf.[16]
- Iraq-Syria Pipeline—This is the export pipeline that the cabinet wanted to use as leverage to get Syrian cooperation.[17] Paul Wolfowitz reversed that guidance in a meeting with General Franks a couple weeks before the war started.
- Other Issues—Guiding principles, legal considerations, and production targets were all covered in the memo.

The memo was declassified in 2013 except for the last section on the Iraq-Syria pipeline. That section was redacted from the original declassification. However, the Syria pipeline slide was fully declassified in 2015 in response to my FOIA request. The page directed the military to "keep the pipeline open only if Syria cooperates on issues of interest to US/Coalition."[18] This direction was recommended by the State Department. As mentioned earlier, Paul Wolfowitz verbally reversed those written orders after they were issued and shortly after Secretary Powell's WMD presentation to the United Nations. He did so during a meeting with General Franks before hostilities started. The main pump station for the pipeline was destroyed in early April 2003. A complete copy of the Feith memo to Rumsfeld is found in my *Lessons Learned* work that was completed in 2006 and published for the public in 2016.

CHAPTER 3

ORHA Start-Up

Lieutenant General Jay Garner Is the First Civilian Leader

Better the Arabs do it tolerably than that you do it perfectly.
—T. E. Lawrence

With the oil sector contingency plan behind us in early January 2003, it was time to coordinate our work with the other prewar planning teams handling electricity and water. Oops! There were none.

In January 2003, the EIPG team was notified that it should prepare for a meeting with retired Lieutenant General Jay Garner. Reports within the Pentagon were that Jay had been selected by Rumsfeld to be the first post-Saddam civilian head of Iraq under the US government.

Garner arrived at our office on the ground floor of the Pentagon and quickly introduced himself. He had a southern accent and a friendly personality. Accompanying Garner were retired Lieutenant General Ron Adams and Larry Hanauer. Adams worked for Garner twice in the 1990s on the army staff at the Pentagon and was now joining Jay as his deputy. Their personalities and styles were different yet complementary. Hanauer was a civilian on Doug Feith's staff assigned to help Jay at the Pentagon but did not go downrange with us.[1]

Mike Mobbs introduced the EIPG team, and we all sat around the large meeting table and reviewed our work of the past couple of months. Garner listened but quickly put on his recruiting hat for his new position. He went around the table asking each of us if we would join him and his team. Everyone gave him about the same answer. They were part of a US government agency and it depended on their specific agency's decision. Seneca mentioned that she would like to, but her husband might have a problem with his pregnant wife entering a war zone.

My reply was different. As an independent contractor, I had no agency ties and had already thought about a deployment while developing the contingency plan. If I was going to be a part of the contingency planning, then I should be prepared to be part of the team to execute the plan. Also, unlike the others on the team, the army could activate me as a lieutenant colonel through my reserve unit. There was really no other choice for me. I looked at Jay and said, "I am not owned by an agency. I'm here for some adventure. If you can promise me some adventure, then I will go."

Jay smiled and replied with his southern drawl that we were going to get a lot of adventure. He said that I was the fourth team member and I should report for his daily meeting the following morning.

The next morning, Garner's meeting started at 0800 hours. Chris Milligan from USAID and Colonel Tom Baltazar were new faces. They increased the team size to six. There were no computers, phones, or other pieces of office equipment necessary to plan for posthostilities work in Iraq. Chris took the only desk that looked like a workstation, since Tom and I already had an office in another part of the Pentagon. The logistical support for standing up a large group was not in place. We met every morning for the next couple of months and new faces were added every day. Office arrangements never did catch up, so many of us maintained our old desks in the Pentagon.

Mike Mobbs, Clarke Turner, and Frosty joined the team within a few weeks and also kept their desks in the EIPG SCIF. Clarke and Frosty joined me on the oil team. Mobbs took the position of director of Civil Administration under Garner after David Kay abruptly quit. Kay stayed in the job a short time and was best known as UN chief weapons inspector following the first Gulf War. Civil Administration was one of three primary groups reporting to Jay Garner. The other two were Humanitarian Assistance and Reconstruction.

The oil group was one of several groups that reported to Mobbs. We had a distinct advantage over the rest of ORHA. We had three people that had just participated in a detailed planning exercise. The other teams were just starting, about three months behind us.

Planning under ORHA took on a tactical approach. More emphasis was placed on specific things that needed to be done in the first thirty, sixty, and ninety days after arrival in Baghdad. Garner labeled these ORHA planning exercises "rock drills." ORHA oil planning started to include more resources from the US Army Corps of Engineers (USACE) headquarters and Major General Carl Strock. Several meetings took place at the Pentagon and the USACE headquarters in DC involving USACE resources.

Major General Strock joined the Garner team as a key leader in Garner's inner circle. The oil group eventually took guidance from Strock just before and during our stay in Kuwait. This proved beneficial to ORHA oil for a couple reasons. First, the initial drive into Iraq was an army operation. Having an army major general helping the oil group provided credibility and support that we would not have received otherwise. Second, Task Force RIO (Restore Iraqi Oil) reported to the CFLCC commander on paper, but the reality was that Major General Strock was the senior USACE officer in theater. Brigadier General Robert Crear was the commander of Task Force RIO, but he was also a USACE general officer. Crear and Task Force RIO took guidance from Strock.

Over the course of about two months at the Pentagon, the ORHA team grew to include about 150 people. Most of the people added were from the DOD. Several came from the Department of State. The remainder came from the other agencies of government.

As the numbers grew, Garner demonstrated his unique ability to recite all the first names at the morning meetings. He made an effort to go around the room after every meeting and ask each individual by name if they had anything to add. Knowing everyone's first name is an admirable leadership trait for any leader but especially for one who would be leading us into a war zone.

The EIPG oil team of Clarke, Frosty, and me morphed into the ORHA oil group during February and March. The key relationships developed during this period were with the ORHA staff and especially Major General Carl Stock, Brigadier General Bob Crear, and the rest of the military officers and civilians of the Corps of Engineers. We continued to coordinate with the CENTCOM staff, even after they moved from Florida to their forward locations in Qatar.

The EIPG team that did not go to Kuwait eventually went back to their old jobs. Seneca went back to the State Department; Barbara Glotfelty and Michael Makovsky returned to their previous work. However, Glotfelty and Makovsky remained engaged in Iraq oil issues at the Pentagon.

One of my more memorable stories during this period was from February 5, 2003, the day Secretary of State Colin Powell made his infamous presentation to the United Nations about Saddam's threat to peace because of weapons of mass destruction (WMD). Everyone in the Pentagon watched the testimony on televisions in their offices. I watched a key part of his presentation in a waiting room outside of Doug Feith's office. I was waiting to see Feith's military assistant, whose office was next to Feith's. I sat for more than thirty minutes, waiting for the general to return. Feith's administrative assistant and I were the only two people in the waiting room outside Feith's office, where we watched Powell's testimony on the television.

Secretary Powell made a strong case that Saddam had mobile biological warfare laboratories located throughout the country, caravans that could be easily relocated and that were therefore hard to detect. At one point, Secretary Powell made a very strong point that the evidence was overwhelming and that something had to be done. When he finished his point, a loud cheer came from several people inside Feith's office. The door was closed, but the noise was quite loud. The administrative assistant and I looked at each other. She commented, "I wonder what they are so happy about." It was obvious that they were responding to the remarks made by Powell. Everyone in the Pentagon watched his presentation that day, but the people in Feith's office were the only ones who seemed to be cheering.

This appeared to be a turning point in the rush to war. The DC area had experienced anthrax attacks after the 9/11 attacks. There was paranoia that something would happen again—more anthrax, a dirty bomb, a mushroom cloud, whatever!

Secretary Powell's case to the UN Security Council was very convincing and hit home for many Americans. An Iraq war did not seem inevitable until hearing the secretary's testimony. Here was a well-known American patriot, a retired four-star army general, a former chairman of the Joint Chiefs of Staff, a highly respected individual within the military, and a public figure who stood for integrity. If Colin Powell said that Iraq was a threat to our security, then it had to be true.

The United Nations presentation did it. The deal was sealed. We were going to war. The only person who could stop us now was a compromising Saddam Hussein, and the chance of that happening was not likely.

In addition to having KBR execute the contingency plan, two additional tasks were assigned to them in early 2003. The first task was to mobilize subject-matter experts to meet with and train the forward elements of the invasion force in Kuwait and teach them how to safely shut down oil equipment. The second task was to procure and position firefighting equipment to fight the anticipated oil well fires. The first Gulf War had several hundred well fires. We were expecting something similar.

Task Force RIO mobilized and moved forward into Kuwait to join the command of Lieutenant General David McKiernan, commander of CFLCC, in early March 2003. Major General Strock and the oil group maintained communications with Task Force RIO from the Pentagon. Operationally, our team was ready to go.

Another administrative issue surfaced before departing the Pentagon. I had been working under a contract through a small engineering firm since October 2002, but that small firm advised me that their contract with the DOD could not handle an international deployment. Garner had me contact the Pentagon personnel office to get a government appointment. The personnel office advised that they were backlogged, still trying to get the appointments for Garner and his team of retired general officers. My appointment would take several weeks to complete and we were scheduled to depart for Kuwait in a week. There were several ORHA team members with this problem. Garner got involved and identified a large contractor, SAIC, to subcontract those ORHA members that needed a method to deploy.

I mention this for two reasons. First, it was another example of the lack of administrative and logistical support we were given at the Pentagon. Second, my quick and high-priority contract solution helped bring about an introduction to a very interesting new acquaintance, Terry Sullivan.

Mike Mobbs introduced Terry to me just three days before our departure. Mobbs told me that Sullivan was a Navy SEAL and he needed to get on the ORHA plane in three days using the same contract mechanism that I had used. Terry was in his mid-fifties and in great physical condition.

"Welcome to ORHA," I said.

Terry responded that he would not be part of ORHA. He was reporting directly to Doug Feith. Sullivan emphasized to me that he needed to get downrange using the same contract mechanism I had used. Terry officially reported to Mike Mobbs, director of Civil Administration under ORHA, and that was how he was deployed.[2] However, what was official and what was reality were not always in alignment.

Terry was initially somewhat cryptic about his job, but he did say that his wife was Carol Haave, deputy assistant secretary of defense for security and information operations. She was a political appointee under Doug Feith.

Terry and I became roommates a couple weeks later at the Hilton in Kuwait, and that was when he shared much more about his mission. His mission was to support Dr. Harold Rhode—keep Harold from getting killed, as Terry said. Harold was a talkative academic type who seemed out of place in a war zone. He had worked in Feith's group at the Pentagon. Harold was described as an ardent Zionist and ideologue who declared that Tehran and Syria were the real objectives.[3]

Terry said that Harold's primary job was to link up with Dr. Ahmed Chalabi's team and support Chalabi's quest to be Iraq's prime minister. In Kuwait, they met with Salem "Sam" Chalabi and others who were part of Dr. Chalabi's organization, the Iraqi National Congress (INC).

Terry's ability to commandeer equipment and navigate the military system was excellent. Unbeknown to ORHA, he organized a land movement of Harold and INC members to Nasiriyah in April 2003 to link up with Dr. Chalabi's first landing within Iraq—a memorable photo op in front of the international press. It was the first time that Jay Garner ever met Chalabi. Garner was surprised to see Sullivan and Harold Rhode in Nasiriyah, escorting Chalabi to meet Garner.[4]

Weeks later in Baghdad, Sullivan and Rhode took on a mission directed from somewhere in the United States to take Judith Miller of the *New York*

Times to help find ancient Jewish artifacts. It quickly became a political, high-priority mission that diverted the military resources of the United States and another country so that it could be completed. The artifacts were secured, frozen in refrigeration units, and flown out of Iraq. Removing these artifacts from an occupied country was in violation of international law and became a serious legal issue between Iraq and the United States. The NSC was still addressing the problem thirteen years later with Iraq.[5]

Terry resigned from his SAIC contract during the summer of 2003 and went to work for Dr. Chalabi's new business ventures in Iraq. He mentioned to me later in 2003 that Dr. Chalabi compensated him with an equity position in a cell phone company contract that Chalabi had just won from the Coalition Provisional Authority (CPA).

The oil planning at the Pentagon went well. Jay Garner became comfortable with our work and the work of many of the other ministry teams. One area that caused him discomfort was governance. How to transition to a new Iraqi government? He met with Doug Feith in early March 2003 to discuss governance. Secretary Rumsfeld had encouraged Garner to discuss such issues with Feith. Garner shared with Feith several of the governance issues that he considered problems. Feith then gave his input. He told Jay that he would make life a lot easier if Jay just appointed Ahmed Chalabi as president when he got to Iraq.[6]

Garner looked at Feith. He could tell that Feith was serious. Garner believed that it was the intention of Secretary Rumsfeld to let the Iraqis choose their own leader. He turned down Feith's suggestion. Feith looked for a positive reaction from Garner but did not pursue it further when Garner rejected his proposal.[7]

A few days later, Garner asked his press officer, Captain Nate Jones, USNR, to put together a press conference. It was just days before our departure to Kuwait and Garner wanted to initiate contact between ORHA and the press before we moved forward. Captain Nate Jones was a newspaper editor from the Midwest when he was not in a military uniform. Jones quickly invited several members of the press to meet with Garner and retired Lieutenant General Jerry Bates. Bates was one of the retired general officers of Garner's ORHA leadership team. The interview went well until the Q and A portion. A quotation from the ORHA historian's work follows and it best describes what happened:

> A seemingly innocent question related to the role of Iraqi expatriates. Garner said a group of Iraqis based in Michigan could probably provide some assistance, but when pressed on details about Ahmed Chalabi's Iraqi Na-

tional Congress, Garner responded, "There is no specific role for the INC, there is no specific role for Chalabi. I don't have a candidate. The degree to which those people play or are allowed to play, we'll determine later."

That night Feith called Garner in a rage. "You've done some real damage to the INC and to Ahmed. . . . You were very non-supportive in your press conference. And by the way, who told you—you could have a press conference?"

"I didn't know I had to ask, number one," Garner responded. "Number two, if we're getting ready to go somewhere we ought to tell everybody what we're doing because the word is going to get out anyway. But the fact of the matter is I'm neither a supporter nor a non-supporter of the INC or Ahmed Chalabi. I don't have a candidate. And by the way, your boss doesn't either. Because I have heard Rumsfeld say several times, 'I don't have a candidate.' And Doug, if this is really bothering you so much . . . then you have a choice. You can say you don't want me and I'll go back to my company. I don't have a problem with that at all and you can put another guy in my job."

Feith backed down, but the next day Garner found he was embargoed from speaking to the press. Wolfowitz also called to express concern but was smoother than Feith: "Maybe we just need to be more careful next time when we refer to the INC, because they've been very supportive." Garner refused to be scolded by Feith, but Nate Jones took the rebuke as a bad omen. Jones anticipated a more draconian response to the interview and he would be proven right.[8]

The last ORHA meeting at the Pentagon was conducted on Friday morning, March 15, 2003. A large room was being used for the daily meetings. Several people stood next to the walls and in the doorway. Colonel Mike Murphy, an air force officer on the White House legal staff, shared some final words of wisdom. Mike was a big Texan whom Jay quickly awarded the nickname Bigfoot.

Bigfoot quoted T. E. Lawrence. "Do not try to do too much with your own hands. Better the Arabs do it tolerably than that you do it perfectly. It is their war, and you are to help them, not to win it for them. Actually, also, under the very odd conditions of Arabia, your practical work will not be as good as, perhaps, you think it is." I would remember those words for the following ten years, and note the ineffectiveness of many arrogant Americans who chose to follow the opposite approach.

The ORHA team consisted of 179 people on March 17, 2003, the day that ORHA departed the Pentagon for Kuwait. We all met in the Pentagon south

Secretary Rumsfeld bids farewell to Jay Garner and the ORHA team in the south parking lot of the Pentagon on March 17, 2003. Photo by author.

parking lot on a Sunday morning to load onto buses for our trip to Andrews Air Force Base. Secretary Rumsfeld came out for some departing words and to bid us farewell.

One new addition was made to the oil team during the day of departure. Clarke Turner had recruited another engineer from the Department of Energy (DOE) on the West Coast, Mike Taylor. Mike was an experienced engineer who had worked with Clarke for several years in the DOE. The oil group was four strong when the ORHA team flew out for Kuwait.

The long flight to Kuwait gave Garner the opportunity to visit all the various ORHA teams for informal updates on our preparation activities. The oil group sat together on the flight.

Garner was concerned about the optics of a recent personnel announcement. Philip Carroll had just been appointed chairman of the advisory board of the Iraqi Ministry of Oil by Secretary Rumsfeld but would remain in DC until May. Garner was concerned about appointing a high-level oil industry executive, the former CEO of Shell US, to the position. Garner felt that such a title was inconsistent with the titles of the rest of his ministry teams. We had to be sensitive to the optics around oil. The press would love to create an oil story where there was none.

The importance of the oil sector to Iraq's economy was well understood. We had to be prepared to quickly respond to the worst-case scenario. It would have been irresponsible to be unprepared for such an important sector. Clarke and I reminded Garner that Carroll's title was consistent with the worst-case scenario we had in our contingency plan. If there was massive destruction of the oil sector, similar to what had happened in Kuwait, and an Iraqi workforce was unwilling to work with us, then we would have to be ready to tap into many resources. Someone of Carroll's stature would be needed, and he would need an appropriate title. Our EIPG plans included the appointment of a chairman and several board members, both Iraqis and non-Iraqis, from the global oil industry. We could adjust those plans accordingly if the outcome of the war was much better than we planned, and we did.

CHAPTER 4

Iraqi Oil Sector

History, Resources, and Organization

Oil has played a role in Iraq's economic history since it was first discovered at Baba Gurgur, just north of Kirkuk, in October 1927. Exports to the Mediterranean started in 1934 through a small pipeline that transported oil to both Haifa and Tripoli. At the time, Haifa was under British control, while Tripoli was under French control. Deliveries to Haifa stopped in 1948, when Iraq ordered the Iraq Petroleum Company (IPC) to stop delivering to the new state of Israel after the first Arab-Israeli War.[1]

IPC had a monopoly concession for all of Iraq's oil production. They produced and transported the oil to their customers in the Mediterranean and Europe while paying small fees to the government of Iraq. The companies currently known as BP, Shell, Total, and ExxonMobil owned IPC. When the shipments to Haifa stopped, IPC needed a new export channel for handling their production from northern Iraq. They constructed a new pipeline to Baniyas, Syria, in 1952 for transporting Iraqi oil to the Mediterranean.[2]

In the early 1950s, Basra Petroleum Company, a wholly owned subsidiary of IPC, discovered giant oil fields in the south and soon started exporting out of the Basra ports. An offshore loading terminal, Khor Al Amaya oil terminal (KAAOT), was constructed in 1959 to export the larger volumes produced from the southern oil fields of Rumaila and Zubair. The higher oil exports brought increased wealth to the country but highlighted to the Iraqis the potential for more wealth and prosperity.

Seven large oil companies, known as the "Seven Sisters," controlled global oil supply prior to the 1970s. Five of those seven had equity positions in IPC. Note that ExxonMobil was two separate companies before 2000. To counter the power of the large oil companies, the major producing countries of Saudi Arabia, Iran, Iraq, Kuwait, and Venezuela met in Baghdad during September 1960 to create a new organization, a cartel known as the Organization of the Petroleum Exporting Countries (OPEC). This marked the beginning of the battle between OPEC countries and the oil companies for a larger share of the oil wealth. Iraq developed an oil policy in the mid-1960s to slowly eliminate IPC. It allowed the newly created Iraq National Oil Company (INOC) to bid on the expropriated fields of IPC. INOC was created in the early 1960s as a state-owned oil company.[3] Disagreements between IPC and the govern-

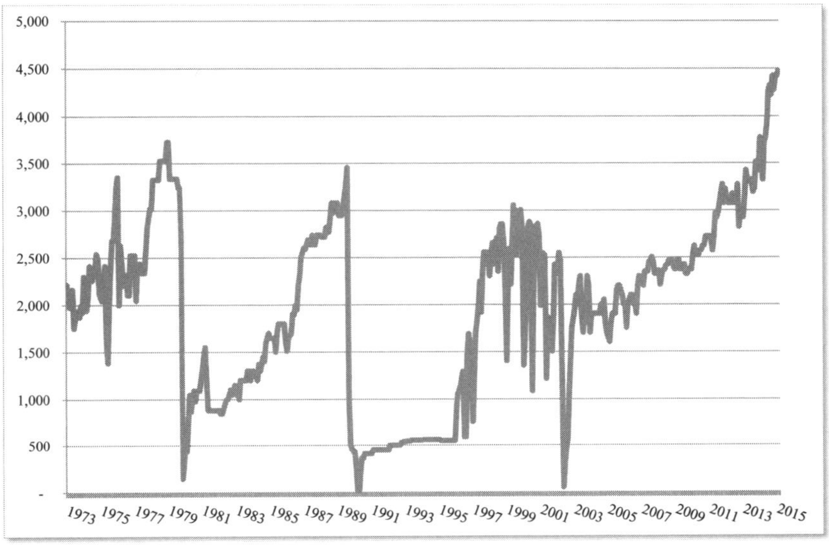

Fig. 4.1 Iraq Oil Production, 1973 to 2015

ment of Iraq about sharing Iraq's oil wealth eventually led to the nationalization of IPC in 1972.[4]

Other OPEC countries had already nationalized by 1972, or would soon do so. Venezuela nationalized in 1976, Saudi Arabia in 1980, and Iran in 1951, followed by another nationalization in 1979, after their revolution.[5] All were motivated by the desire to get a greater share of the profit from the oil companies.

Higher production and export volumes continued through the 1970s and quickly ramped up to more than 3.5 million barrels a day before 1980. The historical volumes of Iraq oil production from 1973 through the end of 2015 are shown (see Figure 4.1).

Higher export volumes boosted revenues into Iraq. The higher oil revenues increased Iraq's reliance on oil for their economy. This led to increased government spending, especially to support the military under Saddam Hussein. The drops in oil production on the graph resulted from the three wars: the Iran-Iraq War (1979), the Gulf War (1990), and the war of 2003.

Oil eventually made up 95 percent of the country's revenues. The US government knew this before 2003. We knew that any attempt to kick-start the economy after liberation had to involve the oil sector. That is why the US government created an oil planning cell in October 2002, months before any planning started for other sectors of Iraq.

Resource Base

As of December 2015, Iraq had the fifth-highest oil reserves in the world (see Table 4.1). They reported oil reserves of 144 billion barrels in 2015, an increase from the 112 billion barrels in 2003. The number should increase over the next twenty years as exploration work identifies more oil. More than 50 percent of the country remains unexplored. Also, existing giant oil fields have experienced higher recorded reserves, primarily as a result of new technology.

Most of the Venezuela and Canadian crude oil in the country reserve table (see Table 4.1) is heavy oil that has a high cost to produce. The Middle East crude oil is lighter and has a much lower cost to produce. The equivalent US conventional reserves (ex-shale oil) are less than forty billion barrels.

Most of Iraq's oil reserves are located in the south, close to Basra. These giant fields are West Qurna, Rumaila, Majnoon, and Az Zubair. The largest field in the north is just outside Kirkuk, and one giant field is located close to Baghdad. Rumaila has been the highest-producing field in Iraq since 2003.

OPEC Production Quota

Iraq traditionally held an OPEC production quota that was equivalent to Iran's. Iran had a quota between three and four million barrels a day from the 1990s to 2010.[6] Saddam's three wars starting in 1979 severely hurt Iraq's ability to maintain their OPEC parity relationship with Iran. Iraq's resource base could support higher production than their historical levels. Their oil field bid auctions of 2009 demonstrated Iraq's potential to sustain a much higher production level than in their past.

The Iraqi oil sector successes since 2003 have changed the landscape within OPEC. The country production table (see Table 4.1) shows the OPEC countries' current production levels versus those in December 2003.

Oil Export Infrastructure

Iraq is essentially a landlocked country with only a small area in the south having access to the Arabian Gulf to transport its oil. Its geography adds a greater complexity to its oil sector compared to other Middle East oil-producing countries. The availability of oil export channels has historically played an important role in the ability of Iraq to produce its oil.

Single-Point Moorings in the South

The largest channel for oil exports at the end of 2015 went through the single-point moorings (SPM) in the south. These SPMs were first installed in 2012. The oil minister described the SPM project from 2012 as the most important infrastructure project in decades during the start-up ceremonies. The US

Table 4.1 Oil Reserves and Production Comparison Data

Country	*Reserves, Billions of Barrels*
Venezuela	298
Saudi Arabia	268
Canada	172
Iran	158
Iraq	144
Kuwait	104
UAE	98

Data Source: 2015 data from EIA

Oil Field	*Discovered*	*Reserves, Billions of Barrels*
West Qurna	1973	43
Rumaila	1953	17
Majnoon	1975	12
Kirkuk	1927	10
East Baghdad	1976	8
Az Zubair	1949	5

Data Source: Wikipedia

Millions of Barrels a Day	*Production, December 2015*	*Production, December 2003*	*% Change*
Saudi Arabia	10.1	8.4	20
Iraq	4.5	1.9	137
Iran	2.9	3.8	-24
UAE	2.9	2.2	32
Kuwait	2.7	2.2	23
Others*	9.3	9.3	0
Total	32.4	27.8	17

*Includes Venezuela, Angola, Nigeria, Algeria, Indonesia, Qatar, Ecuador, and Libya.

Data Source: OPEC Secretariat

government spent $2 million to seed this project, which cost Iraq $2 billion and which should provide hundreds of billions of dollars in revenues over the next twenty years. Chapter 17 discusses all the details about this project from its beginning in 2007.

Al Basra Oil Terminal (ABOT)

This offshore terminal was commissioned in 1975 with four berths and two forty-eight-inch subsea pipelines from shore, about thirty miles to the north. Brown & Root (predecessor of KBR) constructed it with an initial capacity of three million barrels a day. It never operated at that capacity, and the subsea pipelines were de-rated in 1990, decreasing operational capacity to about 1.6 million barrels a day. ABOT is located in the deepest channel in Iraqi waters. The deeper water enables it to load the very large crude carriers (VLCC) capable of carrying two million barrels. Exports loaded before 2003 at ABOT were under the UN Oil-for-Food program.

Khor Al Amaya Oil Terminal (KAAOT)

The smaller of two large offshore oil terminals is located in the northern Arabian Gulf, about thirty miles south of the Iraqi shores. KAAOT is older—it was commissioned in 1959—and is close to the Iranian border. The initial terminal had two thirty-two-inch pipelines that transported oil from the onshore Al Fao oil terminal to KAAOT. These pipelines were eventually replaced with one forty-two-inch pipeline.

The Iran-Iraq War severely damaged this terminal and the forty-two-inch pipeline. The Iranians seized the platform at one part of the war and occupied it until removed by the Iraqis months later. The forty-two-inch pipeline was in poor condition. Satellite photos before 2003 identified oil slicks on the water surface above the pipeline whenever the line was pressured up to load vessels. The terminal was seldom used before 2003, but the estimated loading capacity of the system was about one hundred thousand barrels per day.

Iraq-Turkey Pipeline (to Ceyhan)

This pipeline consists of two parallel lines: a forty-six-inch pipeline commissioned in 1977 and a forty-inch pipeline commissioned in 1988. The initial design capacity was 865,000 barrels per day for the forty-six-inch line and 785,000 barrels per day for the forty-inch line.[7]

There were four pump stations in Iraq and six pump stations in Turkey. One of the Iraqi pump stations located near Mosul was bombed by the coalition air forces during Operation Desert Storm in 1991 and has never been repaired. The capacity on the Iraq side was limited to over seven hundred thousand barrels per day in 2003 because of this damaged pump station, but the northern oil fields seldom produced enough oil between 1991 and 2003 to test this capacity.

The pipeline was the primary export route for Kirkuk crude oil. Iraq maintained storage tanks of about 8.3 million barrels in Ceyhan. Exports through this pipeline were controlled under the UN Oil-for-Food program prior to 2003.

Nonoperational Export Pipelines

The following export pipelines stopped operating either before 2003 or upon the US forces' entry into Iraq. The Haifa pipeline had been closed since 1948, but its status was important to Israeli government leaders and Pentagon civilians in 2003.

Kirkuk-to-Haifa Pipeline (aka Mosul-Haifa Pipeline)

IPC constructed the first export pipeline from Iraq in the early 1930s from the oil fields in Kirkuk. It went through Jordan to the seaport of Haifa in present-day Israel. The Kirkuk oil fields started producing oil in 1927 and this pipeline connected those fields with the Mediterranean.

The pipeline operated between 1934 and 1948, ending after the 1948 Arab-Israeli War, when Iraq refused to ship any more oil to the new state of Israel. Its length was about 585 miles, with a diameter of twelve inches from Kirkuk to the K-3 pump station and as small as eight inches from K-3 to Haifa.

This twin pipeline split at pump station K-3 near Hadeitha, with the second line of twelve inches in diameter carrying oil to Tripoli. Lebanon was under a French mandate at the time. The line to Tripoli was built primarily to satisfy the demands of the French partner in IPC, Compagnie Française des Pétroles.[8]

The Tripoli leg remained operational after Iraq refused to transport oil to Haifa in 1948. It carried oil from Iraq until a new and larger pipeline to Syria was commissioned in 1952.

The Haifa pipeline was not on our radar screen during the prewar planning process. We were looking at operational pipelines. Most of us were not even aware of its previous existence until its status became a high priority during the summer of 2003 by Michael Makovsky in the Office of the Under Secretary of Defense for Policy.

Dr. Ahmed Chalabi had promised to restore the Haifa pipeline during his speeches to prominent Israel lobby groups in the United States during the 1990s. Chalabi knew that a way to get the United States to help him overthrow Saddam was to get the support of the powerful Israel lobby. He would have the oil flowing from Kirkuk to Haifa after the United States helped him overthrow Saddam.[9]

The Haifa pipeline was the subject of several press articles in early 2003. Israel was dependent on oil imports for almost all of its energy needs in 2003. The Israeli infrastructure minister, Joseph Paritzky, publicly spoke in March 2003 of resurrecting the pipeline to Haifa since it could save Israel the high cost of importing oil from Russia. He mentioned that the oil had stopped coming to Haifa in 1948 and had been redirected to Syria.[10]

A couple weeks later, a UK newspaper, the *Guardian*, carried the story "Israel Seeks Pipeline for Iraqi Oil," which discussed Paritzky's comments and the prewar promises of Chalabi. The following quotation from the article identifies the motivation: "Its resurrection would transform economic power in the region, bringing revenue to the new US-dominated Iraq, cutting out Syria, and solving Israel's energy crisis at a stroke." It mentioned the US-sponsored Iraqi National Congress and its leader Ahmed Chalabi, the former banker favored by the Pentagon for a powerful role in the war's aftermath. It quoted a former senior CIA official: "It has long been a dream of a powerful section of the people now driving this administration [of President George W. Bush] and the war in Iraq to safeguard Israel's energy supply as well as that of the United States."[11]

Minister Paritzky's comments were not the only public announcements coming from Israel at the time. The Israeli finance minister, Benjamin Netanyahu, said he expected an oil pipeline from Iraq to Israel to be reopened in the near future after being closed when Israel became a state in 1948. "It won't be long when you will see Iraqi oil flowing to Haifa," the port city in northern Israel, Netanyahu told a group of British investors, declining to give a timetable.[12]

These press clippings did not reach our oil team in Baghdad or the US military commanders trying to maintain stability in Iraq in the summer of 2003. However, they became well known by the Iraqis. The unintended consequences of these public Israeli announcements contributed to serious security problems for our military forces in Iraq. They will be discussed later in the book.

Kirkuk-to-Baniyas Pipeline

This was a crude oil pipeline from the Kirkuk oil fields to the Syrian port of Baniyas. The pipeline was about 552 miles long, and the initial capacity was three hundred thousand barrels per day. It consisted of twin pipelines, one thirty inches and another thirty-two inches in diameter, which opened on April 23, 1952.

In 1950, the Iraq Petroleum Company contracted Bechtel to construct the twin pipeline. Iraq nationalized IPC in 1972, and this led to the nationaliza-

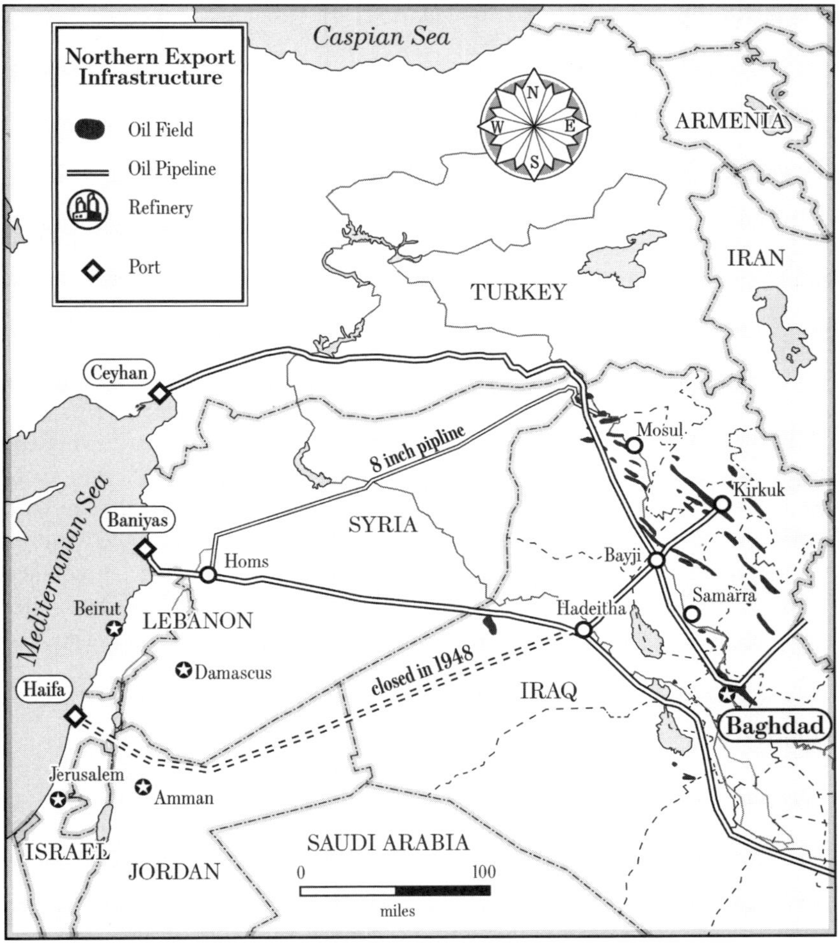

Northern Export Infrastructure as of 2003.

tion of IPC's assets in Syria, including the Syrian section of the Kirkuk-Baniyas pipeline.[13]

In November 2000, numerous press reports indicated that Syria and Iraq had reopened the Kirkuk-Baniyas pipeline. The Middle East Economic Digest (MEED) reported initial deliveries of Iraqi crude oil through the line at around 140,000 to 150,000 barrels per day.

The mission of the 75th Rangers was to ensure that oil stopped flowing through this pipeline upon our entry into Iraq. The main pump station for this pipeline was located close to Hadeitha in Anbar Province. The pump station was destroyed in early April 2003.

Sufaya-to-Syria Pipeline

This is a small pipeline in Ninewa Province, northwest of Mosul. It was constructed in 1985 to transport crude oil from the Ayn Zalah oil field to Syria.[14] The pipeline eventually connected Syrian oil fields in eastern Syria to Homs.

The pipeline was eight inches in diameter at the border crossing and had an initial maximum capacity of fifteen thousand barrels per day. It had been operating at a rate of about fifteen hundred barrels per day just prior to hostilities in March 2003. The crude oil was bartered with Syria to get about forty megawatts of electricity for the citizens of Ninewa. The pipeline was shut down during hostilities and Syria stopped sending the electricity.

This pipeline became a source of controversy in Washington during the summer of 2003. In the middle of the hot summer, the pipeline was again used to transport oil to Syria in exchange for badly needed electricity and petroleum products. Washington's desire to penalize Syria had to be weighed against the negative impact to the Iraqi people and the casualties being taken by the US military. As I will discuss in later chapters, some of the louder voices in DC assigned a lot of value to penalizing Syria while giving little or no value to the suffering Iraqi people and the lives of US soldiers.

Iraq Pipeline through Saudi Arabia (IPSA)

This pipeline was commissioned in 1988 and was used to export oil from the southern fields across the Arabian Peninsula to the Red Sea port of Al Murajiz, just south of the Saudi industrial city of Yanbu. The pipeline was a forty-eight-inch pipeline with an initial capacity of 1.6 million barrels per day. The pipeline operated for only a couple of years, until Saddam invaded Kuwait in August 1990. Saudi Arabia confiscated the pipeline in 2001 as compensation for Iraq's war debts to the kingdom.[15]

Refining

Iraq had three large refineries in 2003 in Baghdad (Doura), Bayji, and Basra. Another eleven small refining units of about ten thousand barrels per day were located throughout the country. One was an underground refinery in the hills close to the Bayji Refinery that never operated. Saddam thought that such a refinery was a good way to protect refining capacity from the enemy. It was dismantled after 2003 and the parts were used elsewhere.

The Doura Refinery was constructed in the 1950s, the Basra Refinery in the 1970s, and the Bayji Refinery in the early 1980s. All three used old technology and Bayji was the only refinery with a cracking unit. Cracking technology enables a refinery to produce higher yields of gasoline and diesel products from a barrel of crude oil by cracking the longer molecules of the heavy oils.

The refining sector had excess capacity for producing diesel products. Iraq had exported this excess volume in the past. However, it barely had enough gasoline capacity to meet domestic demand in 2003. The imports of new vehicles during 2003 and 2004 increased gasoline demand, and imports had to fill that gap.

Organization

Most oil-producing countries have a regulating function that falls under a ministry of oil and an operating function that falls under a state-owned oil company. A good example is Saudi Arabia (KSA). KSA has the Ministry of Petroleum, which handles the regulating functions. Aramco is the national oil company that handles oil operations, which is owned by KSA.

Prior to 1989, the Iraqi structure was very similar. There were the Ministry of Oil and the Iraqi National Oil Company (INOC). This changed in 1989, when both the regulating and the operating functions were moved under the Ministry of Oil (see Figure 4.3).

The Ministry of Oil employed about seventy thousand Iraqis in early 2003. Most of these employees worked for one of the fifteen state-owned operating companies under the ministry. These companies were set up as independent companies, each with their own board of directors and a director general in charge. However, the director general reported to a deputy minister and the deputies reported to the minister of oil.

The largest company was the South Oil Company (SOC), headquartered in Basra. It had responsibility for all oil production and exports in the south, and had more than sixteen thousand employees. The North Oil Company (NOC) had ten thousand employees and had the same responsibilities in the north. Its headquarters was in Kirkuk.

The South Refining Company (SRC) was in Basra, and the Basra Refinery had a capacity of 160,000 barrels a day. SRC had small refining units in Maysan and Nasyriah. The Midlands Refining Company (MRC) was in Baghdad, and its Doura Refinery had a capacity of 110,000 barrels per day. MRC had small units in Najef, Diwanyiah, and Samawah. The North Refining Company (NRC) was in Bayji, and its Bayji complex had a capacity of 310,000 barrels a day. NRC had small units in Hadeitha, Kirkuk, Kisak, Qayyarah, and Saniyah.

The South Gas Company (SGC) was located southwest of Basra, in Khor au Zubair, with a plant in the North Rumaila oil field. The North Gas Company (NGC) was located in Kirkuk.

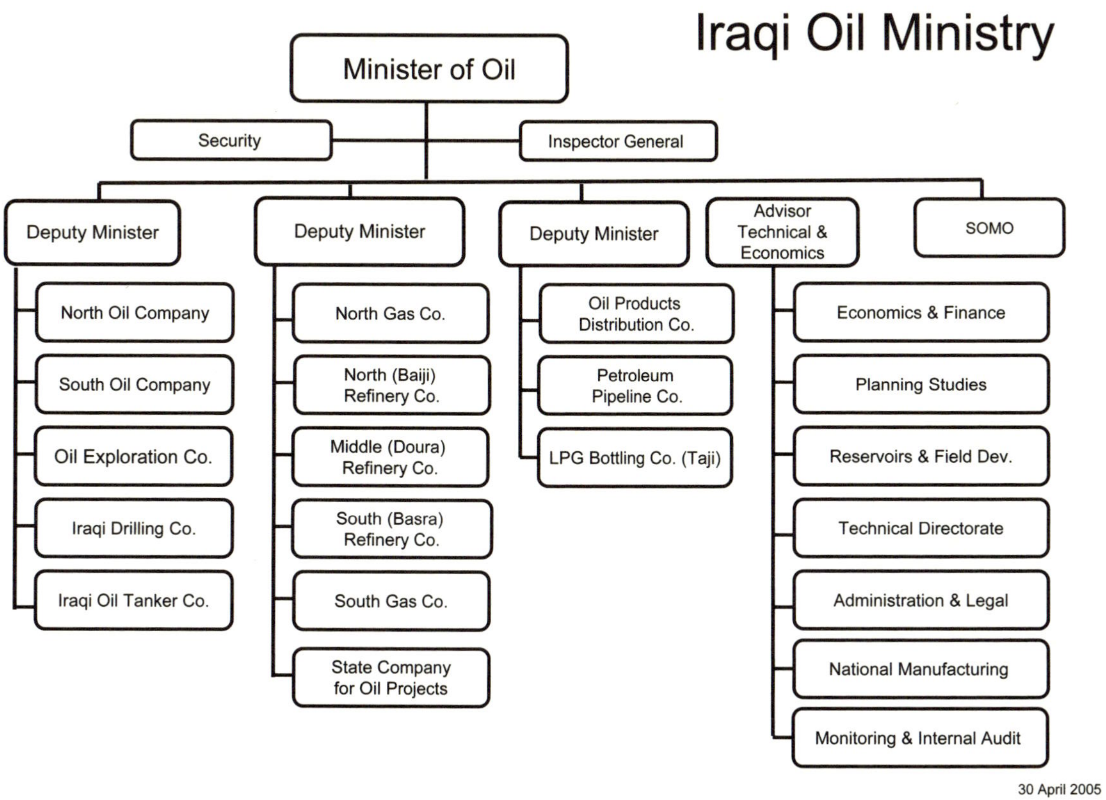

Fig. 4.3 Organizational Chart of Iraqi Oil Ministry

The headquarters for the Distribution Company, Product Pipeline Company, LPG Bottling Company, Drilling Company, Exploration Company, and State Company for Oil Projects were all in Baghdad.

Besides the operating companies, the ministry consisted of several directorates with staff functions, such as Economics & Finance, Planning, Technical, and Administration. A director general was in charge of each of these directorates and reported directly to the minister, or to the senior deputy minister. All of the staff directorates were located in the Baghdad headquarters.

The State Oil Marketing Organization (SOMO) reported directly to the minister of oil. It had about 150 employees in 2003 and was responsible for marketing and exporting oil for Iraq. A director general led SOMO. It was physically located in Baghdad, about a mile from the ministry headquarters.

The organizational structures below the director general level were a function of the type of company or staff directorate. For example, a refining or gas company had several managers who reported to the director general: production, maintenance, technical, or offsite plant managers. The types of managers in a staff directorate were normally unique to that staff function.

Several changes took place in the oil ministry organizational structure after 2005. Here are the most significant:

- The addition of the Petroleum Contracting & Licensing Directorate (PCLD) before the oil field auctions of 2009: This group administered the auctions and oil field development service contracts with the international oil companies.
- The addition of Maysan Oil Company and Dhi Qar Oil Company in southern Iraq: The increased oil production in the provinces of Maysan and Dhi Qar qualified those provinces to have their own oil companies, similar to South Oil Company.

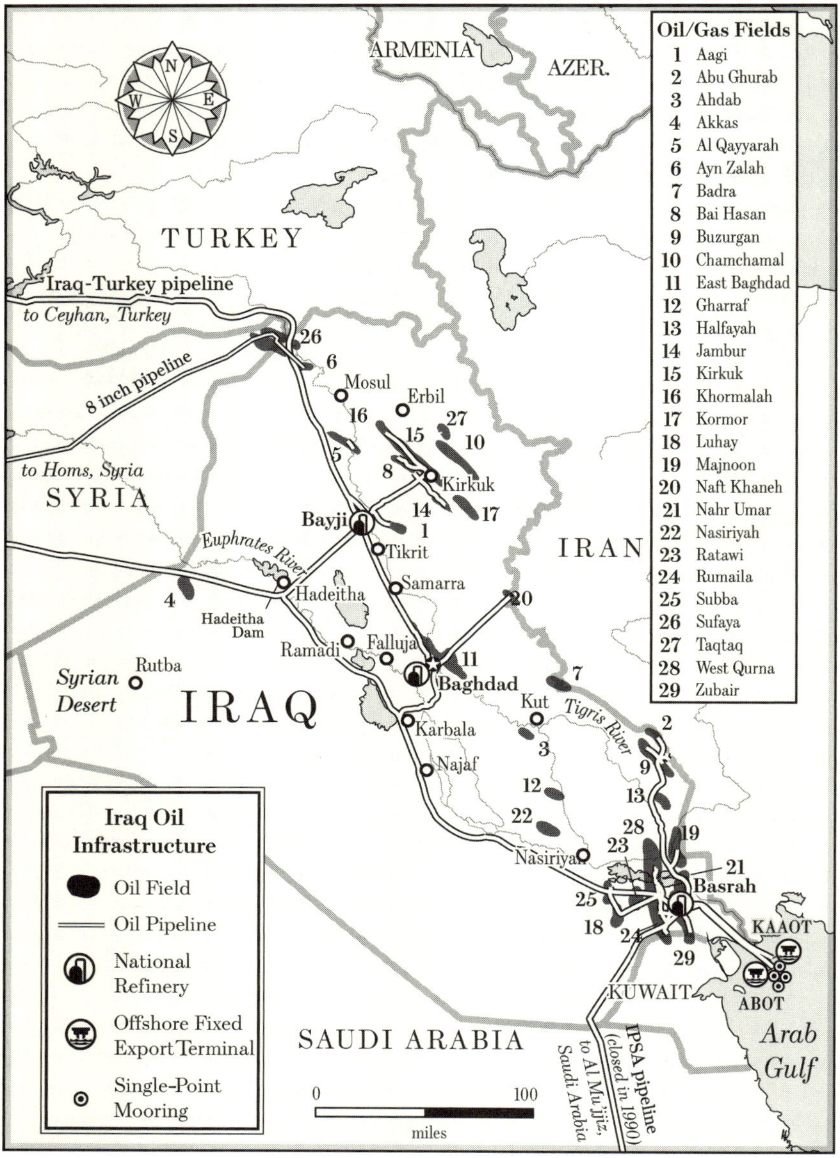

Iraq Oil Infrastructure as of 2015.

CHAPTER 5

Adjusting the Oil Plan

We lost a Marine CH-46—all twelve on board were killed.
—Marine Lieutenant Colonel Paul Shelton

We landed in Kuwait on March 18, 2003. Our home for the next month was at the Hilton Kuwait Resort, located along the Arabian Gulf beach on the south side of Kuwait City. It was a gorgeous resort that had recently been completed. The hotel had a couple hundred rooms, plus several four- to six-bedroom cottages. The cottages proved to be functional in assisting with ORHA planning and preparation.

The first couple of days were highlighted by rocket attacks coming from southern Iraq. We quickly grabbed our protective gas masks and moved to assigned shelters in the basement of the Hilton. This happened several times during our first couple of nights.

On the third day, March 20, Ron Adams approached me about joining him and Ambassador Robin Raphel for a trip to Qatar, to the CENTCOM forward headquarters. Garner needed to improve his connection with the CENTCOM headquarters in Qatar to better achieve unity of effort.[1]

We caught a C-12 military aircraft for the trip. The C-12 is a Beechcraft twin-engine turboprop used for transporting small groups. We were advised that morning that the war had started.

Arriving at the US air force base in Qatar, we were informed of bits and pieces from the press as to how the war was progressing. Upon our arrival at the CENTCOM headquarters, we were escorted into a conference room to meet with General Abizaid and many on the CENTCOM staff.

As we moved into the conference room, Lieutenant Colonel Paul Shelton quickly approached to give me an update on the oil sector. Paul calmly stated that things were going quite well. The ground war kicked off because it appeared that Saddam was trying to start several well fires and destroy the oil fields, similar to what he did in Kuwait in 1991. Shelton and Cliff Fowler both confirmed that there were only seven fires burning at the time. The leading combat elements had secured all of the oil fields located west of Basra and were moving north to Nasiriya.

When Paul advised of this huge oil sector success, much better than the five hundred well fires we were expecting, my mind began to race through the entire Contingency Support Plan (CSP). Our focus for the first six to nine months was supposed to be fighting well fires. We had pre-positioned firefighting equipment in Kuwait and had brought in teams from Boots & Coots Inc., Red Adair's original oil well firefighting company. Additional equipment outside of Kuwait was identified and scheduled for delivery. Very little of this would be needed now. These fires would be put out in a few weeks. Our focus needed to quickly turn to getting the oil flowing again. I was thinking to myself, what were some of the things I needed to do to get the oil flowing?

Then, I suddenly realized that everyone in the room was looking at me. Ron Adams had asked me a question, twice. By now we were sitting in a large meeting room with about forty people surrounding us. I looked around and realized that my brain was elsewhere, even though I was physically in this meeting.

"Ron, I'm very sorry, but I did not hear your question," I said as I stood up. I do not know to this day the question that Ron asked.

I looked around the room and my eyes focused on Shelton sitting against the wall in the back. "I am not sure that everyone here understands the significance of what Paul Shelton has just told me. If you have successfully moved through and secured the southern oil fields, then you have accomplished something extremely important, and you have put the reconstruction of the oil sector six to nine months ahead of schedule. I don't know how to express my exhilaration, excitement, happiness."

Shelton stood up and cut me off before I could say much more. "Gary, we did not tell you the bad news. A helicopter transporting several troops to secure the Al Fao oil terminal south of Basra went down. We lost a Marine CH-46; all twelve on board were killed."

Upon hearing this news, I fell back into my seat. The realities of this war hit hard. It was as if someone punched me in the gut. This was just the first of many similar gut punches over the next nine years. With this dose of reality, we got back to our meeting.

We spent a couple of days at the CENTCOM headquarters and observed some of their technology. One such high-tech tool was the blue force tracker system in the intelligence section. Cliff Fowler gave me a thorough and technical tour of this latest high-technology equipment. On the large, flat panel television screens were maps that indicated the blue forces (coalition). The blue forces were moving north at the time and the red forces (Iraqis) were resisting on the north. GPS devices were placed on many of the key vehicles of the US forces. Cliff could drill down on the map so that I could see the actual movement of vehicles toward Nasiriya.

"Cliff, I understand the use of GPS on our vehicles so that you have the location of our units, tanks, and trucks to the nearest meter. How do you know where the Iraqis are located, the red forces on your screen?" I asked.

Other visitors had asked the same question of Cliff Fowler during previous tours so he was ready with a humorous response. He said that as much as he would like to put GPS devices on the Iraqi tanks, he had not been successful with convincing the Iraqis to allow him. We all got a chuckle over his comment. Identifying the enemy forces on the map was the work of a lot of intelligence resources throughout the theater, coupled with analysis by well-trained soldiers. They were able to place Saddam's specific military units and their hardware.

A lot of time was spent with Paul Shelton's J-5 (plans) colleagues, including his supervisors Army Colonel Michael Fitzgerald and Air Force Major General Gene Renuart. The J-3 (operations) staff was also working on oil issues. Colonel David Halverson was the contact in operations, along with Lieutenant Colonel John Agoglia. John would eventually become the CENTCOM liaison to our offices in Baghdad, so he would see us quite often over the following months.

Army Lieutenant Colonel Wayne Dudding, a US Army Reserve infantry officer, was assigned to the J-2 (intelligence) staff, but his civilian profession became important in the next couple of weeks. Wayne was an engineer who worked for a chemical company in plants back in the Pittsburgh area. Colonel Halverson, in a couple of weeks, would insert him into a high-priority ranger operation in Anbar Province to use his engineering skills.

Meetings were scheduled with my West Point classmate General John Abizaid. John was the deputy CENTCOM commander and would soon become the commander in a few months, when General Tommy Franks retired.

Our team returned to Kuwait City after a few days. The huge successes of the initial days of the war forced us to make significant adjustments to our war plan. As General Eisenhower said, "War plans are useless, but the planning process is indispensable."

Task Force RIO was collocated with the Coalition Forces Land Component Command (CFLCC) headquarters at Camp Doha, on the north side of Kuwait City. It was about an hour's drive from our hotel. We made several trips there in the following week to coordinate with Brigadier General Crear and his RIO team. By this time, he had added several reserve officers from the Texas and Louisiana areas who worked for oil and chemical companies along the US Gulf Coast. His program manager was an experienced and competent USACE civilian by the name of Gary Loew. Gary would eventually play a key role in developing our master plan in a couple of months.

My first day in Iraq was April 7, 2003. Major General Carl Strock and the ORHA oil team moved forward into southern Iraq, home of the Rumaila supergiant oil field, where well fires were still burning. General Crear and a few of his RIO folks joined us. Crear also brought along about a half-dozen members of the press.

The well fires could be seen from many miles away. Our convoy approached one of the blazing well fires and parked our vehicles about a hundred meters away. The intense heat from the blaze prevented us from getting much closer. Employees of Boots & Coots were on the scene. They were subcontracted under the RIO/KBR task order at the time. Also at that location were employees of the Kuwait Oil Company, Kuwait's large national oil company. The Kuwaitis were well trained at putting out well fires since they had extinguished several hundred fires in Kuwait in the early 1990s.

One of the most impressive firefighters was a young female engineer with the Kuwait Oil Company, Sarah Akbar. Sarah and her Kuwaiti firefighters put out most of the seven fires in southern Iraq, a very impressive accomplishment.

Our preparation work back at the Hilton Hotel with the ORHA team and our oil group continued. We linked up with Brigadier General Jack Kern, commander of the 352nd Civil Affairs Command. A notable officer under General Kern was Colonel Bob Schroeder, who would join me in Baghdad in a couple of weeks. Also, Colonel Bryan Hanes would remain in Kuwait for his entire tour in 2003 and be our primary contact with South Oil Company. Bryan was an employee for Schlumberger, an oil field service company in Oklahoma, when he was not wearing his military uniform. He and I would eventually commute from Kuwait City to Basra every day for my last week in Kuwait before I left for Baghdad. His oil experience coupled with his hard work made him extremely valuable during those early weeks.

After our visit to the oil fields in southern Iraq, Jay Garner wanted us to quickly make contact with the oil ministry leadership, located in the Basra area. We did not have any knowledge of the key leaders at that point, but we did know the location of some of the oil facilities. We organized a convoy of several SUVs, and joining us were General Crear and part of his RIO team.

About a dozen of us took off early one morning headed to the Basra Refinery, located about ten miles west of the city of Basra and a three-hour drive from our hotel. It was a memorable journey. Once we entered southern Iraq, Iraqis came out to the road and waved to us as we passed by. We saw destroyed Iraqi military equipment still smoldering from battles that had taken place within the last few days. We saw Iraqi looters, with material that they had taken from various government facilities, walking along the road with their donkey carts loaded to the max.

Our convoy pulled into the Basra Refinery and parked next to the process units. I jumped out of my vehicle and noticed a British officer running toward me. He was shouting, "Who are you? What are you doing here?" He was a young British lieutenant and his platoon was responsible for securing the refinery. They had just moved into this location early that morning and were setting up their defensive perimeter. The lieutenant was a bit animated and irritated with us, announcing that there was still hostile fire going on in the area. Then Crear exited his vehicle and the officer quickly calmed down. He thought we were just a convoy of lost civilians in his area of operation until he saw an American general officer.

There were a few young Iraqi engineers close by, but their English-language skills were poor. We had a difficult time communicating. One identified himself as a production engineer and said that he would go get the refinery manager for us. About fifteen minutes later, a gentleman in his early forties with a patch over his eye walked toward us. It was Ibrahim Thaer, the technical manager and not the refinery manager. Thaer's English-language skills were much better than the other Iraqis'. We asked where the refinery manager was located, and he said that the British soldiers had incarcerated the refinery manager and his deputy. Needless to say, Thaer was not happy about being the top guy at that point, since the two previous top guys had been taken off to prison.

Thaer gave us a refinery tour. The refinery processed 160,000 barrels per day of crude oil at maximum capacity and produced gasoline, diesel, kerosene, and heavy fuel oil. It was old technology with low yields of high-value products. For example, it yielded about 47 percent gasoline and diesel products from Basra crude oil, while a US Gulf Coast refinery would yield more than 90 percent from the same crude oil.

Thaer mentioned that he and several of his engineers had to secure the refinery the previous day before the British troops had arrived because the looters tried to steal everything that was not tied down. The tour of the main control room was memorable. The equipment was several decades old and used old analog control technology. The military foxhole that had been dug just outside the refinery control room caught our attention. No refinery operator in the West would consider a foxhole inside the refinery process units as something necessary.

Thaer showed us around the entire refinery that day. He shared more information, as he got comfortable with us. Much of my oil industry experience with Mobil Oil was in the refinery sector. All of his process units were quite familiar to me. Clarke Turner asked him about the patch over his eye—if that was something he had received during the recent attack. Thaer said he had been fighting eye cancer for a couple of years and he was optimistic on this day that his treatment had been successful. Unfortunately, the cancer returned and became fatal about five years later.

After making this initial contact at the Basra Refinery, we continued to make contact with the various Ministry of Oil elements in the south. Our US Treasury team wanted to make payroll for the workers in the oil fields. It seemed like a good idea. ORHA wanted to encourage the workers to return to work. Paying them $20, more than a month's normal pay, would be a payment of good faith and gain their trust to return to work. This turned out to be confusing to the workers. Receiving money was not urgent for them. KBR was assigned the task of making the payments to the various workers. Having KBR make these payments gave some people the impression that they were now working for KBR and not the Ministry of Oil.

The British forces appointed a leader of oil operations in the south in April, Jabbar Al Luaibi. Jabbar had been the number three or four person in South Oil Company prior to hostilities. The previous director of South Oil Company, a young Baathist from Baghdad, had driven to Baghdad before hostilities and had not returned. The second-in-command refused to leave his home. Jabbar was selected and appointed by the UK forces to take responsibility for all oil operations in the Basra area. He let it be known through the UK forces that KBR should not be paying his Iraqi employees. He requested a meeting with the Americans at the Red Cross/Red Crescent offices in downtown Basra.

When word came that a senior leader of the oil ministry wanted to meet, we immediately responded. I made contact with Pam Quanrud at the NSC in the White House. Pam advised me that Jabbar was well known and had been an excellent choice by the British forces. He came from a respected Basra family and had been educated in the United Kingdom. Pam quickly sent talking points for me to cover during the meeting. These points were the standard US message from our "Guiding Principles" developed during EIPG. They included that Iraq's oil belonged to the Iraqi people and that the United States was not there to steal the oil. Pam's forethought was good; she was always looking for ways she could add value to our work in the field.

We met in downtown Basra at the Red Crescent offices. On one side of the table were Jabbar, his deputy Kifah Numan, and a few other individuals who were trusted leaders within South Oil Company. On our side of the table were Clarke Turner, General Crear, a couple of people from Task Force RIO, and myself. After introductions, I thanked Jabbar and the Red Crescent for hosting this important meeting and then mentioned the talking points submitted by the NSC.

Jabbar quickly got past the issue of the United States not being there to steal the oil. He wanted to immediately focus on the payroll payments being made by KBR. His point was that by having KBR pay his employees, we were confusing them as to whom they worked for, unless our intent was to have his people report to KBR. This was his test. Were we going to rely on his leadership or were we going to take control? I quickly assured him that it was not our intent for KBR or the US military to control the oil sector. I explained that the ORHA leadership was concerned that the employees might not have enough money while things were in disarray. The payments were made strictly as a measure of goodwill to convince them to go back to work.

He must have believed me. Otherwise, he would have walked out of the meeting. I certainly would have if I were in his shoes. Paying employees is extremely important in the employer/employee relationship, as anyone working in the private sector certainly understands. This understanding was not as well appreciated within the US government. This lack of appreciation became a very serious issue to our mission later that summer.

We quickly moved on to the subject of how to get the oil flowing again in the south. Jabbar shared with us some of the difficulties he had been experiencing with getting equipment returned that had been looted over the last couple of weeks. General Crear assured Jabbar that RIO was prepared to supply material to help bring production back online in the oil fields.

The meeting terminated after about forty-five minutes and we agreed to meet again within a few days so that Jabbar could bring a list of items that were urgently needed by South Oil Company. This was the first of several meetings that occurred over the next few weeks. RIO and KBR supported Jabbar in his goal to quickly restore production in the south.

As we left the Red Crescent offices in Basra, I noticed Chip Cummins of the *Wall Street Journal* outside. I had met Chip just a few days earlier during our first press day with ORHA. Chip had reported in the *Wall Street Journal* that a former ExxonMobil executive was a senior lead on ORHA. Chip's article was the first notice to my former colleagues at ExxonMobil of my new job.

"What are you doing here? I understand there is still shooting going on in the area. Hopefully you have security with you?" I asked him.

Chip smiled and replied that he heard there was going to be a story. He asked if I had any comments for the press.

"Please get your quotes from the Iraqis. Jabbar Al Luaibi was the person we met with today from South Oil Company. And I hope you get back to Kuwait safely," I shouted as I walked toward my SUV.

I would see Chip many times over the following couple of months in Baghdad, whenever there was an oil story. Chip and his fellow reporters in the press braved many dangerous places to get their stories. My appreciation of their bravery grew over the next several years.

Our next meeting with Jabbar and several of his leaders took place a few days later. We brought KBR with us. This meeting and those that followed every few days were very productive. Jabbar brought a list of several dozen items that were urgently needed. General Crear referred to these lists as Jabbar's Christmas lists. Doug Fletcher and his fellow engineers from KBR reviewed the lists with Jabbar during our meetings. After understanding the material and the urgency, KBR made telephone calls in the middle of the meeting to estimate when the material could be delivered.

These meetings produced results at an early stage. A year later over dinner, Jabbar and I reflected on these early times as the most productive. The support provided from the US government to South Oil Company was excellent during the first six months. Results were quick. Unfortunately, this would change before the end of 2003.

One of our bigger challenges was to locate heavy equipment: backhoes, cranes, and other machinery to assist with installation of the material being received. The heavy equipment had been looted during the first few weeks, and Jabbar had to approach the looters to get some returned. He told many stories of requesting the return of this material while looking down the barrel of an AK-47 in the hands of the looter. His South Oil Company offices had been looted and burned. He organized his people to return to work and to rebuild the offices, only to have them looted again a month later.

The year 2003 was extremely difficult for Jabbar and his leadership team at South Oil Company. The US government was fortunate to have worked with such a capable and driven person in charge of South Oil Company during this very difficult period.

The meetings with Jabbar continued every few days. Colonel Bryan Hanes and I commuted daily from Kuwait City to Basra to meet with Jabbar and others. We worked to facilitate his projects with RIO, as well as any management issues where our help was required.

Iraq's communications network was destroyed during hostilities. A key service provided to the Iraqis was the issuance of satellite telephones for all the oil leaders in the Basra area. These satellite telephones were difficult to use but better than nothing. One had to be standing outside with an unobstructed view of the sky in order to make contact with the satellite. Until cell phone service was provided in the Basra area later in 2003, the satellite phones served for communications within Basra and eventually with the oil ministry leadership in Baghdad.

Our team in ORHA had been augmented in mid-April and a new face appeared on April 17. John Kjar, an Australian government employee, had been assigned to our team, and would play an important role in Baghdad working with the SOMO group. Four new Department of Energy employees arrived on April 21: Gary Holcomb, Julio Maldonado, David Callahan, and Lee McLemore.

Also on April 21, Jay Garner deployed to Baghdad. Garner called a meeting of his leadership team and the ministry senior advisers the night before his departure. Garner spoke of the need for himself and his advance party to be physically present in Baghdad as soon as possible.[2]

He covered several issues that evening. The most important was his empowerment talk. Garner mentioned that when we got to Baghdad and worked with our ministries, we would need to make decisions. He was empowering us to make decisions, as needed. Garner told us the only wrong decision was NO decision. He went on to say that he had our backs. If anyone wanted to criticize us for any decisions we made, then the criticism would be of Jay Garner. I think it was safe to say that every senior ministry advisor felt empowered.

One rising star on the ORHA team was a young State Department GS-13, twenty-nine-year-old Sherri Kraham. "Sherri was attractive, articulate, politically savvy, and ambitious."[3] She was appointed as a staffer under Colonel Paul Hughes, who headed the strategic planning group. One of Sherri's responsibilities was keeping track of key policy decisions. When Jay was finished talking that night, Sherri commented that there were two important decisions that Doug Feith's office was supposed to make before we left the Pentagon. She said that nothing had yet been received from Feith, but the senior advisors could not wait any longer. Guidance on the two decisions was needed immediately. One was a De-Baathification Policy, and the other was the Iraqi military.

Garner shared the following De-Baathification Policy. He said that when we got to our ministries in Baghdad, the bad Baathists would have been either killed or run off. The remaining Baathists in leadership positions were probably pretty good. He told us to work with them, but, more important, to see if the Iraqis would work with them. If the Iraqis accepted them as their leaders, then we would accept them. If the Iraqis did not accept them, then we had to replace them.

What to do about Baath Party members in the ministries was an important issue, similar to the decision made after World War II in regard to Nazi Party members in the postwar German government. Garner's De-Baathification Policy seemed like a practical approach to take. It was apparent that Jay had given this a lot of thought. This approach with the Iraqis appeared to be a good start.

Concerning the Iraqi army, Jay always took the position that the army would be used to rebuild the country. They would not be disbanded.

Ambassador Bremer would issue two important and dissimilar policy decisions in writing a few weeks later on De-Baathification and the Iraqi military. He signed the decrees and took responsibility for both policies, but the origins of these controversial decisions were in Feith's group at the Pentagon.[4]

Ambassador Bremer shared with me in December 2015 that Doug Feith handed the two controversial decisions to him the day before he left for Iraq. Feith told him that the two policies were being sent out to Iraq for Jay Garner to implement. Bremer volunteered to take them with him and implement them once he arrived in Baghdad. He was comfortable that his two decrees were consistent with other work previously accomplished within the US government.[5]

We had our normal meeting with the Iraqis on April 23 at the Basra Refinery. At the end of the meeting, Jabbar and I had a private discussion. We discussed my movement into Baghdad the next morning. It was my intent to link up with his colleagues in Baghdad and to work closely with the Iraqi leadership in Baghdad, as we had in Basra. Jabbar gave me a boost of confidence when he said that his colleagues at the Ministry of Oil would welcome me as he had welcomed me a couple weeks before.

An important date was April 23, 2003—first oil. The Iraqis shut down 90 percent of production prior to March 20, 2003. The remainder was stopped by noon on March 20, when hostilities started.[6] April 23 was the first day that the oil started flowing again from the oil fields, and it has not stopped since. South Oil Company, RIO, ORHA, and the press met at a large gas-oil separation plant (GOSP) in the Rumaila oil field. The oil started flowing from the oil wells at high pressures and through high-pressure flow lines to the GOSP. Production quickly grew to over fifty thousand barrels per day, enough to supply the Basra Refinery and power plants in the south.

The press took many pictures and video clips that day. One video continues to be played on news channels in the United States whenever a discussion of oil takes place. It is a video of oil gushing up from an oil pond. It appears that oil is just spouting out of the ground to fill the pond. Actually, one of the high-pressure flow lines had failed, and it took a while to get someone to fix the break. The pipe fracture occurred just outside of the GOSP, and a pond of oil quickly appeared, with a six-foot gusher in the middle of the pond. I am reminded of that important date of "first oil" whenever I see the video clip on television. Thank you, CNBC.

CHAPTER 6

Penalizing Syria

Why So Important?

The United States was moving to cut off an important oil pipeline running from Iraq into Syria.
—Secretary Donald Rumsfeld

One of our initial planning goals was to minimize any damage to the oil infrastructure during hostilities. The military did a good job by kicking off the war early and securing the southern oil fields before hundreds of well fires could occur. There were only seven well fires and they were extinguished quickly. In April 2003, it looked like the damage from the war would be minimal and the cost to the US taxpayer would be confined to the damaged oil wells.

One late change to our original plan was a high-priority mission to eliminate the flow of oil to Syria from Iraq on the Iraq-Syria pipeline. I attended the video teleconference (VTC) in March 2003 while still at the Pentagon. Paul Wolfowitz gave General Franks very explicit instructions that no oil should ever flow through the Iraq-Syria pipeline after we entered Iraq. These verbal directives reversed the written instructions included in an action memo from Doug Feith to Secretary Rumsfeld on January 16, 2003. Those written instructions specifically directed the military to "keep the pipeline open only if Syria cooperates on issues of interest to US/Coalition."[1]

The mission was assigned to the 75th Ranger Regiment. The rangers were inserted into Anbar Province early in the war with the mission of seizing and protecting the Hadeitha Dam. Several firefights with Saddam's forces occurred before the rangers secured the dam and then they moved to secure the K-3 pump station, the main pump for the Iraq-Syria pipeline.

As previously mentioned, Lieutenant Colonel Wayne Dudding, USAR, was working on the CENTCOM J-2 intelligence staff in Qatar when it was recognized that the oil mission required some additional expertise to ensure that the flow of oil through the pipeline to Syria ceased. Dudding worked for a chemical company in Pennsylvania and the Northern Panhandle of West Virginia when not wearing a military uniform, so he got the assignment as a subject-matter expert (SME).

Dudding caught a C-130 flight into Iraq in early April, landing at Al Asad air base, aka H1. From H1, he took a memorable night Blackhawk ride, flying at very low altitude across the beautiful Lake Qadisiyah on a dark night with a sky full of stars and no moon. This manmade lake had over sixty miles of shoreline and supplied irrigation water to Anbar Province. Damming the Euphrates River north of Hadeitha formed Qadisiyah. The flight was under tactical conditions since multiple firefights in the area were still occurring. The rangers had just recently secured the dam.[2]

Dudding landed at the dam on the same day that the 3rd Infantry Division made their initial thunder runs into Baghdad. He linked up with the rangers in preparation for early morning movements to the K-3 pump station.

The weather was much colder in Hadeitha than in Qatar, and Dudding had to depend on his chemical protective suit to keep him warm that night. He did not sleep well his first night in Iraq. Desert dogs located nearby were howling all night, keeping him awake. Dudding assumed that they were enjoying their spoils of war with the remains of the dead Iraqi soldiers still lying in the desert.

Dudding and his ranger team left at daybreak for K-3. Upon reaching the pump station, an Iraqi soldier came out of a building spraying bullets toward the convoy. He was quickly neutralized. The rangers had been in the area the previous day and had returned to the dam at the end of the day. Dudding saw that the housing structure enclosing the large gas turbine pumps was severely damaged and engulfed in flames. The heat from the fire was intense and had been burning for several hours before his arrival. Dudding said that his original mission was to help the rangers shut down the pump station. His mission quickly changed from shutting down the pump station to helping the rangers extinguish the fire.[3]

The inferno was being fed by the backflow of oil being gravity-fed from the storage tanks a hundred meters away as well as the oil in miles of pipeline downstream of the pump. Dudding recognized that they needed to find the valves in the pipelines to stop the flow of oil to the fire. The local Iraqi manager had to be found.

Standing at the fence line waving a white handkerchief was a middle-aged Iraqi that the rangers had been ignoring. Dudding heard the radio traffic of the rangers talking about this persistent Iraqi at the fence. He told them to bring the man to him. It was the pump station manager.

The Iraqi manager showed Dudding a flow diagram of the pump station in his office. He identified the valves needing to be closed. The Iraqi called in his firefighting team and together with the rangers was able to extinguish the fire and cool down the pump station. The damage was severe. The building and several gas turbine pumps were ruined. There was no way that the pump facility could be repaired. It would require a complete rebuild.

Dudding and the Iraqi manager talked after the fire was extinguished. Wayne remembered the Iraqi's teary eyes when he told the story of the bombing of the station in 1991. The Iraqi manager shared with Wayne that it took several years to rebuild his pump station after the first bombing in 1991. He had only started pumping again to Syria a couple of years before. The pump station manager was just doing his assigned job.

Dudding received an urgent call from the battalion commander back at the Hadeitha Dam. He needed Wayne to stop everything he was doing and focus on this very important question: Did he think the oil flow to Syria was shut off?

Wayne replied that there was no way any amount of oil could flow through that pipeline with the destruction of the pump station. When Wayne returned to the dam later that day, the battalion commander showed him the unclassified message that prompted the question on the radio a few hours earlier. It stated that the president wanted to know if oil was shut off to Syria.

Dudding spent a couple of days with the rangers before returning to CENTCOM. On his return trip, he had a young soldier approach him in a waiting room at the airfield in Qatar. "Sir, why are you so dirty? What have you been doing?" Dudding was still wearing the same filthy uniform he wore while fighting the fire. He did not take any extras for the mission and he had not taken a shower. Wayne explained that he had been on a mission in Iraq and began showing the soldier some of the video he had taken while extinguishing the pump station fire. The soldier looked up at the television in the waiting room to see the exact same video footage being shown on CNN.[4]

Like many reserve officers I worked with in Iraq, Lieutenant Colonel Wayne Dudding was able to use his civilian expertise to provide a valuable service to the oil mission. He was awarded a Bronze Star two years later, in June 2005, for his work at pump station K-3 from April 9 to April 15, 2003. The award states, "His on-site technical expertise for 3rd Battalion, 75th Ranger Regiment, Hadeitha, Iraq, supported the President's strategic objective."

Besides CNN, the story was on the front page of the *New York Times* on April 16, 2003. At the Pentagon, Defense Secretary Donald H. Rumsfeld said the United States was moving to cut off an important oil pipeline running from Iraq into Syria, the first major economic penalty against Syria.[5]

Unfortunately, this penalty would eventually penalize the people of Iraq much more than Syria. Iraq lost an export channel and some very expensive gas turbine pumps. Syria lost only the future tolling fees from those exports.

The Iraqis first advised me of the K-3 attack in May 2003, but I did not appreciate the level of destruction until seeing the pictures and hearing the $50 million rebuild estimate developed by the Task Force RIO contractor in June 2003. This destruction of the large pumps at the K-3 pump station was the most effective way for the rangers to complete the mission that Paul Wolfowitz gave to General Franks during the March VTC.

Two pieces of oil infrastructure were bombed during the first few weeks of hostilities: Al Fatha Bridge and the K-3 pump station. The first was unintentional, a mistake. My oil contacts at the Defense Intelligence Agency (DIA) explained that the intelligence community did not realize the bridge had several oil pipelines under the road that crossed the Tigris River. The bridge was targeted to prevent the movement of Saddam's military during hostilities. The target was reviewed and approved by the intelligence groups. The oil pipelines were unknown to those responsible for approving the bombing targets. Our oil mission was to repair and restore the oil infrastructure to prewar levels. It took three years and several million dollars, but all the oil pipelines crossing the Tigris River were eventually repaired.

The K-3 pump station was a well-known piece of oil infrastructure. It was marked on all of our maps and the oil tanks were visible to any pilot dropping their ordnance. Our repair and restore mission did not apply to the K-3 pump station. That mission applied to all oil infrastructures except for the K-3 pump station. In June 2003, we identified all oil infrastructure damaged during the war and estimated the cost to repair that damage. Task Force RIO estimated that the K-3 rebuild would cost $50 million. Our orders from the Pentagon civilian leadership were explicit. The US government would not rebuild K-3, but the Iraqis could use the $50 million to build anything else in the oil sector the Iraqis wanted.

From 2006 through 2011, US and British senior military officers would visit the K-3 pump station, where the mangled metal from the ruin still existed. The steel I-beams that supported the roof of the pump station building were still warped from the tremendous heat experienced during April 2003. The gas turbine pumps were rusting metal junk. After seeing the destruction, some would ask me why we had to bomb the pump station instead of just shutting it down and closing the valves. My response surprised some: "Short of being able to read the minds of our political leadership at the time, I will probably never understand why."

It was during my research for this book that I discovered a credible and logical explanation. As previously mentioned, the history of the Kirkuk-to-Haifa pipeline may help shed some light on additional reasons for the decision to destroy the K-3 pump station. The Haifa pipeline operated between 1935 and 1948, ending after the 1948 Arab-Israeli War, when Iraq refused to

Gas turbine pumps and housing at the K-3 pump station destroyed in April 2003. Photo taken by Colonel Paul Dietrich USAR during a visit in 2006. Photo by Dietrich.

ship any more oil to the new state of Israel. To replace the Haifa pipeline, the Iraq Petroleum Company (IPC) contracted Bechtel to construct a pipeline to Syria in the early 1950s.[6] A key component of the Iraq-Syria pipeline was the K-3 pump station. The export pipeline to Syria would directly compete with a future rebuild and expansion effort of the Haifa pipeline.

As previously mentioned in chapter 4, a UK newspaper, the *Guardian*, carried a story, "Israel Seeks Pipeline for Iraqi Oil," which discussed Israeli minister Paritzky's comments about opening the Haifa pipeline and the prewar promises of Dr. Ahmed Chalabi to rebuild it. The article identifies the motivation: "Its resurrection would transform economic power in the region, bringing revenue to the new US-dominated Iraq, cutting out Syria and solving Israel's energy crisis at a stroke." It mentioned the US-sponsored Iraqi National Congress and its leader Ahmed Chalabi, the former banker favored by the Pentagon for a powerful role in the war's aftermath. It quoted a former senior CIA official: "It has long been a dream of a powerful section of the people now driving this administration [of President George W. Bush] and the war in Iraq to safeguard Israel's energy supply as well as that of the United States."[7]

CHAPTER 7

Finding Gold in Baghdad

Oil Ministry Leadership

So you know what it's like to be gobbled up.
—Unknown oil ministry director general

It was April 24, 2003, and I landed at the Baghdad airport along with seventeen of my ORHA colleagues, all senior advisers to Iraqi ministries. The flight manifest included Ambassadors Robin Raphel, Tim Carney, and David Dunford. There was also the US Treasury team of David Nummy, George Mullinax, and Van Jorstad. Lee Shatz of Iran hostage fame was the Department of Agriculture representative. The C-130 aircraft pulled up to the terminal labeled as the VIP terminal. As we walked inside the building, a tall female air force lieutenant shouted to us, "Welcome to Baghdad. The pallet on your right contains bottles of water. The pallet on the left contains MREs. The restrooms are Porta-Johns located out back. Make yourselves at home. You are going to be here a couple hours."

Our flight from Kuwait had been delayed, so the military police escort left on another mission and had to be recalled to meet us. The terminal was busy with not just US military and civilians but several nongovernment organizations (NGOs) from other countries. They were busy bringing in people on aircraft. There were several NGOs from Saudi Arabia and Arabian Gulf countries. The air force officer had her hands full trying to keep the terminal organized. Her priorities included providing plenty of water and meals ready-to-eat (MREs).

Over the next eight years, I would travel through this terminal many times. Significant improvements were made with each visit and my last would be September 2011. The creature comforts available during my final trip were significantly better than during my first trip to Baghdad. This terminal was also the site for the final farewell ceremony for the departure of US forces from Iraq in December 2011.

The military police arrived late in the afternoon and loaded us into their armored vehicles for the thirty-minute trip to the Green Zone, the area inside Baghdad secured by the US military where we were based.

Halfway to the Green Zone, I noticed the Doura power plant off in the distance. Only one of the four smokestacks was emitting smoke. Brigadier General Steve Hawkins of the Corps of Engineers had been in Baghdad during the prior week trying to turn on the electricity. How many weeks would it take him to get smoke coming out of all four stacks?

My expectations became lower over the next several years as the difficulties of reconstruction emerged. On every trip to and from the airport over the next eight years I always checked to see how many smokestacks were operating. Seldom was it higher than two. During my final trip out of the country in 2011, two smoke stacks were operational.

Our convoy pulled into the front of the Republican Palace, my new home for the next seventeen months. The four large busts of Saddam on the top of the palace dominated the skyline. They were very large, grotesque images of the former dictator. These busts would be removed within the next year.

Entering the palace front doors, I noticed the dirt and darkness inside the rotunda. Our fellow ORHA colleagues that had arrived a couple of days before were there to greet us. An advance party of less than a hundred folks, led by retired Lieutenant General Jerry Bates, had convoyed up from Kuwait. We were the first element of the main body that would be arriving over the next few weeks. All of the ministry advisors were taken to the south wing of the palace. We were assigned offices on the first floor and living quarters on the second floor.

Walking down the dark halls to the south wing, I noticed that all of the windows in the palace had been shattered by the bombing campaign conducted a few weeks before our arrival. Also, the Iraqis had removed all of the doors from the rooms to store them in a secure place to prevent looting. The ORHA advance party found the doors and was in the process of reinstalling them.

Carrying my large duffel bag over my shoulder and my army cot at my side, I was shown to my room at the end of the second-floor hallway. There were items of furniture still in the room from the previous inhabitants. The mattress on the bed was filthy and the rest of the furniture was gaudy. There was glass on the floor next to the windows, so I erected my army cot just inside the door. There was a bathroom attached to my room, but there were no lights and no running water. The stench was so bad that I had to keep the bathroom door shut to confine the odor. The other rooms on the floor were not much better. Some of my colleagues brought out the candles and flashlights. I took the opportunity to get a good night's sleep. It was my first night to sleep in a palace.

• • •

My first priority was to link up with Colonel Joe (Bob) Schroeder of the 352nd Civil Affairs unit. Bob had convoyed up from Kuwait the previous day and was now probably somewhere inside the Green Zone. Bob had the vehicles and the soldiers to enable me to perform my mission. It took several hours, but we eventually linked up late in the day.

Communications were very poor during those early times. My Thuraya satellite phone was ineffective for contacting anyone inside of Iraq. Both parties had to be outside with clear access to a satellite in order to communicate. It was great for phoning anyone outside Iraq, but it took some coordination to phone inside the country.

My satellite phone was put to excellent use during my second day in Baghdad. The smell of barbecue smoke spread throughout the area at noontime. I walked to the back of the palace in search of the aroma. KBR was cooking hamburgers and serving to anyone that wanted lunch. Many soldiers were in line to get their burgers. I was tired of MREs and was hungry so I joined them.

The young soldier in front of me was talking to his buddy about home and their families. Both soldiers could not have been much more than nineteen years old. I asked when was the last time they had spoken to their parents and they both replied that it had been a couple of months. One said that his mother was probably very worried about him. He said that she worried all the time.

"Where does she live?" I asked.

"My parents live in northern Florida," he replied.

"It is about 4 AM in Florida now. Would she be upset if you woke her up?" I asked.

"What do you mean, sir?" he asked.

"I have this satellite telephone here and I cannot seem to use it to talk to anyone here in Iraq, but it is great for phoning the United States. Let me show you what I mean. What is your mom's telephone number?"

The young soldier smiled and gave me the phone number. I waited until I heard a ringing tone and then handed it to him.

"Mom, mom it's me. Yes, it's really me. I'm okay. Serious, I am okay. I can't tell you where I am. Yes, seriously I am okay. Doing great."

Their conversation continued for another ten or fifteen minutes, until it was time for the soldier to get his hamburger. He signed off by saying, "Goodbye. Love you too mom."

He returned the phone to me and said that his mom was surprised. That was one of the most gratifying calls made with that satellite telephone!

Colonel Bob Schroeder and I met early the next morning in front of the palace for my first trip to the Ministry of Oil. We had two Humvees and two assigned drivers. The Humvees had canvas roofs and did not have any doors or armor. They were strictly a means of transportation. The drivers and Colonel Schroeder had M-16 rifles. I was not armed.

Bob pulled out a map of Baghdad and laid it on the hood of the vehicle. The location of the Ministry of Oil was different than it was on the maps that we were shown while in Kuwait. Our destination was labeled as the Ministry of Defense on his map. The defense ministry had used the oil ministry complex until 2001 while their new facilities were being constructed. The oil ministry moved back to their complex a couple of years before our arrival, but our maps did not reflect the change.

We took a direct route out of the Green Zone across the Marne Bridge and directly down what the military called Route Irish. The 3rd Infantry Division painted their Marne Division emblem on the bridge and renamed it in April to the Marne Bridge. The Iraqi name was the Al Jumariyah Bridge. There was very little traffic during that first trip and no traffic lights, so we made it in about ten minutes.

Our first stop at the ministry was with the US military unit from the 3rd Infantry Division. The young captain gave us a short tour, telling us that one of his biggest problems was searching the employees returning to work. One of the senior Iraqi oil leaders had requested that any ministry employee wanting to return to work be allowed to enter the complex. To accommodate this request without compromising security, every employee had to be searched. The captain said that more female soldiers were needed to search the female employees.

The 3rd Infantry Division was actually the second US military unit to secure the oil ministry complex. The US marines were first. The marines came into Baghdad from the east on Route Irish. They were the first US military unit on the route. Their mission was to stop at the Ministry of Sports/Olympics. They were to search that ministry and look for documents that could be used against the former regime. The complex was engulfed in flames when the marines arrived. The heat was intense, so they moved to their second objective a couple blocks to the west, the Ministry of Oil. They searched that complex for documents.

The oil ministry complex was an excellent base for them. There was a thick wall around the perimeter, twelve feet high, and sentry lookouts at various locations along the wall. There were only a few entrances and they could be easily secured. The building was nine stories high. It was the highest in the area, so that soldiers could be stationed on the roof and watch in all directions. Bottles of gasoline with stuffed rags were found in the complex, but the marines had arrived before anyone could ignite the gasoline. The marines were given the order to hold their position until properly relieved. The 3rd Infantry Division provided that relief.

• • •

Critics of the Bush Administration highlighted the fact that our military protected the oil ministry complex while the rest of Baghdad was looted and burned. They mentioned this to support a position that the war was all about oil. There was a stronger argument; the reason it was not looted and burned came down to military tactics. The complex was a fort inside of Baghdad: a secure position that any company commander would covet. It provided the highest position in the area for observation. It had a high wall for the unit commander to secure his position and protect his soldiers. The looting and burning continued throughout Baghdad for weeks after the marines first entered the oil ministry complex. Only those facilities that had a full-time presence of US forces were spared from the looters and arsonists.

Colonel Paul Hughes and Mike Ayoub of ORHA found a top secret document at the Iraqi Ministry of Defense dated January 23, 2003. It was an emergency plan anticipating the fall of Baghdad. It identified the procedures to be taken by Saddam loyalists after the fall. At the top of the list was to loot and burn all the government buildings.[1]

Several oil ministry facilities throughout Iraq were looted and burned. Even the headquarters in Baghdad had some level of looting before the marines arrived. The SOMO headquarters was located a short distance from the ministry headquarters. SOMO had responsibility for selling all of Iraq's oil. That facility was severely looted and burned. The oil ministry research facility was also nearby. It contained the geological records for the oil reservoirs in the country. The geological core samples for all the oil fields were stored in aluminum containers. That facility was looted. The core samples were dumped out so that the looters could take the aluminum containers.

We entered through the front of the oil ministry complex. There was an area just inside the front door of the headquarters used for displaying pictures and models of the destruction to the oil sector from the Gulf War in 1991. Some of the captions next to the pictures were in English and were very critical of the US government of George H. W. Bush in 1991. The destruction to all three of the large Iraqi refineries by our air force was significant. It was amazing to me that those three refineries could be reconstructed. However, the refineries started operating within a year and were rebuilt within two years.

Colonel Schroeder and I were escorted to the fourth floor of the ministry headquarters, where we found that about a dozen of the ministry leaders had assembled in a conference room. I was not sure if they were expecting me since there were no secretaries to announce us.

Entering the room, I said, "Excuse me, sorry for the interruption, my name is Gary Vogler and I am General Jay Garner's senior oil advisor." I made a point of walking around the room and greeting each person, shaking their hand, looking them in the eye, and repeating each name.

"We understand that you are an Exxon employee," one Iraqi said.

I stopped, smiled, and looked at him, "I was a Mobil Oil employee for nineteen years and then Exxon purchased us. I stayed with ExxonMobil for two years and left the company over a year ago."

"So you know what it's like to be gobbled up," another Iraqi said.

I laughed. "I suppose I do. I will have to share your analogy with my old Mobil friends. Big difference, we don't plan to be here long," I said. "The Exxon gobble was permanent."

"How long will you be here?" someone else quickly asked.

I turned to the person asking that question. It was a serious question and one that deserved a serious answer.

"My job is to help you get your oil production back to a level where you were producing a couple months ago. The sooner you accomplish that, the sooner I will get out of your way and go home," I said.

The senior person in the room spoke up. "Please Mr. Vogler, come with me to my office so that we may talk." He was Deputy Minister Mazen, the senior deputy minister of oil. Dr. Masen had just been appointed to the position about six months before, in late 2002. He had previously been the president at the University of Technology in Baghdad and was a high-level member of the Baath Party. He was one of many competent Baathists that would be removed from the government. Dr. Mazen was quite personable, articulate, and knowledgeable about the oil sector. We chatted about many subjects. He described well the pride of Iraqis in their heritage.

He traced their civilization back five thousand years. The area of present-day Iraq was called Mesopotamia and considered to be the cradle of civilization. Sumer and the Akkadian, Babylonian, and Assyrian empires were native to this area of modern-day Iraq. The former college professor gave me a history lesson while we chatted.

He mentioned that while his ancestors were living in civilized societies three thousand to five thousand years ago, my ancestors in Europe were still going through an uncivilized savage era of continuously slaughtering one another. The subtle affront did not go unnoticed.

I smiled and replied, "So, you are telling me that my ancestors were quite similar to your recent dictator, Saddam Hussein." We continued to chat and get to know each other.

It was a good first meeting with Dr. Mazen, but I left his office troubled. I needed a plan to help these gentlemen get organized, to start working again. There were currently too many unknowns, too many questions, and I needed answers and advice from my NSC contact.

• • •

The return trip from the oil ministry that day was memorable. Our two Humvees were moving through the market area in the center of town, which the American troops had labeled the "looters market." The market was full of street vendors that were peddling the items that had just been looted. Sergeant Angela Leverette drove my Humvee, trailing Colonel Schroeder's vehicle.

We were stopped near the looters market when gunshots were heard. Hundreds of people panicked and ran past my vehicle, bouncing off the hood and bumping into my right arm. There were no doors on our vehicle so I felt vulnerable. Sergeant Leverette grabbed her M-16 rifle, pushed the safety switch from the safe to the semiautomatic position, and put her finger on the trigger. She looked at me and asked me if I had a weapon.

"No, sergeant, my life is in your hands." I smiled to try to assure her that things were OK, but I am sure that I was not convincing.

There were no additional gunshots and the crowds eventually dispersed so that we could continue our journey back to the Green Zone. I learned later that one of the street vendors shot at some thieves trying to steal his goods.

My conversations with Pam Quanrud at the NSC were daily for the first couple of weeks; sometimes there were several phone calls a day. That evening she had good news for me. She had arranged for me to meet three oil leaders from the ministry the next day at the Civic Center. Her intelligence had identified these as the "best and the brightest" inside the ministry and people I should utilize. In hindsight, twelve years later, the three were excellent choices, three gold nuggets that I found in Baghdad: Thamir Ghadhban, Dr. Radhwan Al Saadi, and Dr. Faleh Al Khayat.

Thamir Ghadhban was born in Al Hindyia, a city on the Euphrates halfway between Hilla and Kerbala in 1945. He was educated in the United Kingdom at University College and Imperial College in London, receiving an MSc in petroleum reservoir engineering. He started his oil career in 1973 with the Basra Petroleum Company (BPC) in the southern oil fields.

Ghadhban joined the Ministry of Oil in 1989 as the director general of reservoir and field development in Baghdad. He was appointed as director general of the Planning and Studies Directorate in 1991. In 1994, he was detained and removed from his director general position for criticizing Saddam's spending on palaces rather than other needs of the country. He was released after a couple of months but could not travel abroad for years. He was assigned as the technical advisor to the oil minister.

Ghadhban used this postdetention punishment period to refine his engineering, communication, and computer skills while working on several special projects for the oil minister. He liked to boast to me that he became quite proficient during his penalty period with the Microsoft Office suite of Excel, Word, and PowerPoint. He was quite capable with Microsoft, but I eventually learned to appreciate his leadership talents as superior to any I had previously experienced in the oil industry. He won back his old job of director general of planning in 2002 and held that position upon our arrival.

Thamir Ghadhban and I would become good friends over the long hot summer, not unlike two soldiers that shared a foxhole during wartime. I learned to depend on him and I believe he counted on me during some difficult times. We remain good friends today.

Dr. Radhwan Al Saadi was educated in the United Kingdom. He held a master's and PhD in geology from the University of Wales. He was the director general of the Economics and Finance Directorate, a portfolio similar to that of a chief financial officer of any integrated oil company.

Al Saadi was also an impressive oil executive who had excellent communication and decision-making skills. I quickly gained respect for his keen intellect and recommended him during the CPA era to support other non–oil sector initiatives such as the Trade Bank of Iraq and letters of credit initiatives undertaken by CPA. We considered him an excellent replacement for Ghadhban if something ever happened to Thamir. Radhwan became a good friend and still remains a friend today.

Radhwan was the younger brother of Dr. Amer Al Saadi, an Iraqi scientist who was well known in the international scientific community. Amer had the prewar job of convincing the international community that Saddam did not have WMD. Dr. Amer Al Saadi appeared on international television after Secretary Powell made his now infamous presentation to the United Nations. Amer denied the allegations made by Secretary Powell as a "typical American show, full of stunts and special effects." For his attempt to stop the invasion, Dr. Amer was identified by the US as a high-value official in Saddam's regime. He was number thirty-two of the top high-value political targets, and the only non–Baath Party member in the top fifty. He spent more than two years in US detention after turning himself in to the marines in early April 2003. Several of us attempted to seek his release after hearing the facts behind his story. Unfortunately, it took much too long.

Dr. Faleh Al Khayat was educated in the United Kingdom. He spent more than thirty years in the ministry and was the director general of the Technical Directorate in 2003. He was very knowledgeable about all aspects of Iraq's oil sector and could articulate the technical details better than most. I learned to seek and respect his opinion on all issues during the early days. Dr. Faleh had been a member of the Baath Party, albeit not a high-ranking member. In fact, he told me more than once that he worked hard to come up with reasons to avoid attending meetings. His Baath Party membership eventually caught up to him in late 2003 when the Iraqis purged the ministry of the remaining former members of the Baath Party. His termination by the Iraqi De-Baathification committee was reckless and irresponsible. I labeled it throwing the baby out with the bathwater.

The ORHA and CPA oil teams would work with many high-caliber Iraqi oil executives in the ministry during the early months. Many were Western-educated with excellent English-language and decision-making skills. They were professional oil executives skilled in the oil business. Many were excellent leaders in their field, but the three executives identified by the NSC and our intelligence community truly were the best and the brightest.

April 27, 2003, proved to be the most productive day of my first week in Baghdad. Colonel Bob Schroeder and I met with the three Iraqis: Ghadhban, Al Saadi, and Al Khayat. It was a constructive two-hour meeting at the Civic Center, as described in chapter 1. Our focus was strictly professional. We selected leaders for each of the senior positions based on demonstrated performance and leadership potential. Baath Party affiliation was not part of the equation, positive or negative.

At one point in the meeting, Dr. Faleh Al Khayat asked about the US position on Baath Party membership. His words were direct, "Would we consider Iraqis guilty by association?"

"My guidance from retired General Jay Garner was that no one should be excluded from consideration, unless they were guilty of crimes or participated in programs involving weapons of mass destruction," I replied.

I did not realize that this guidance would have such a short shelf life when I said it. A new De-Baathification Policy would reverse Garner's guidance within a month. Dr. Ahmed Chalabi would lead a committee of Iraqis in a few months that would purge the ministries of all the Baathists.

The first post-Saddam Ministry of Oil leadership team was identified. We selected all the leaders for key positions except for one, the top job. We agreed that a meeting would be held at the ministry headquarters on May 3 to announce the appointment of the entire leadership team, including the person to fill the top job. The deputy minister positions would not be filled until 2004.

My phone call that evening with Pam Quanrud at the NSC included an update about the afternoon meeting, and Pam volunteered to lead the vetting of candidates for the top job. The three Iraqis I met that day were considered for the job, but Dr. Faleh Al Khayat was a member of the Baath Party and Dr. Radhwan Al Saadi was the brother of Dr. Amer Al Saadi, who, as mentioned earlier, had criticized Secretary Powell's WMD presentation at the United

Nations. I mentioned that Thamir Ghadhban was the director general of the planning directorate and that most oil company CEOs spent time in the corporate planning function. Ghadhban was at the top of our short list, if he passed the vetting process.

While standing in front of the palace that evening talking to Pam on my satellite phone, I heard a loud burst of gunfire, followed by .50-caliber tracer rounds flying above my head. They seemed close, but I suspect that they were several hundred feet above me. The sun was down and dusk was turning into darkness, so it was easy to detect the tracers above me. We quickly ended the call and I sought cover back inside the palace. It was just another reminder that I was still in a war zone.

Quanrud and her team completed the vetting process over the next few days and notified me of their decision late on May 1. She advised me that the final decision to select Thamir Ghadhban was agreed on at the Deputies Committee meeting. I immediately tried to contact Ghadhban on the satellite phone, but he did not have his satellite phone in receive mode, outside, and in a direct line of sight to a satellite.

The next morning, I started phoning him every thirty minutes. I needed to confirm that he would accept the job that we were offering. Finally, he answered at about 6 PM and confirmed that he would accept the position. We agreed to meet at the ministry early the next morning, prior to the 10 AM general meeting with the new leadership team.

I pulled into the oil ministry parking lot at 8 AM on May 3 and noticed that Ghadhban was arriving at the same time. We met in the parking lot and walked to the front door together. The US military was searching each employee as they entered the building.

As we approached the military inspection station, Ghadhban noticed three ladies standing off to the left of us. One was crying and the other two were visibly concerned as well. Thamir looked at me and said, "Gary, I need a few minutes here."

He went over and spoke to the three ladies for what seemed to be several minutes. He spoke to them in a reassuring voice. I remember thinking to myself, "This guy will be accepting the senior position within his ministry in less than two hours, speaking to his leadership team and then to the international press. He has a very busy and important two hours in front of him. The time to prepare was very limited and he was taking some of that time to comfort these three female employees. I think I like this guy, and should be able work with him."

The incident with the three female employees was the first of many examples where I observed a quality in Ghadhban that contributed to my respect for him.

Our meeting was held at 10 AM, as scheduled, in the large conference room on the ground floor of the oil ministry headquarters. The oil ministry leadership team was present, including outgoing Senior Deputy Minister Mazen. ORHA's British General Tim Cross gave the opening speech. Clarke Turner had arrived from Kuwait the previous day and was also in attendance.

General Cross and the outgoing senior deputy minister made brief comments. I followed and my comments were also brief. First, I recognized the work of two Iraqis: the directors general of the Basra and Doura refineries for leading their engineers in fighting back the looters and saving their refineries from destruction over the previous weeks. Then I introduced the new chief executive officer (CEO) of the Ministry of Oil, Mr. Thamir Ghadhban.

Thamir gave a very supportive speech to his new team, highlighting the need to get the oil sector operational as soon as possible. He commented that no country likes to be invaded and occupied. He initially hesitated accepting the position for that reason, but he felt that all Iraqi leaders needed to step forward and accept responsibility for the sake of the Iraqi people. He appealed to everyone's sense of nationalism and challenged everyone in the room to join him.

We took a brief recess from the meeting for thirty minutes to address the international press that had assembled just outside the conference room in the hallway. Thamir was the only person speaking to the press. He was now the Iraqi face of the oil sector. I stood off to the side and remained silent. It was Ghadhban's show!

Returning to the conference room, Thamir and I took questions from the leadership team. There was interest in the role of Phil Carroll and me. I highlighted that Phil would be arriving in another week or two and I did not want to speak for him, but both Clarke Turner and I would be working with the ministry in a support role to help the new CEO.

Thamir requested that he be given time with his team after questions so that he could conduct a meeting in Arabic. As Clarke and I left, I noticed that Thamir moved to the head of the conference table in a symbolic gesture. He was now in charge.

The appointment letters given to each director general on May 3 proved vital. This letter served an important role during the first few weeks, controlling the challenges to the professional leadership of the Ministry of Oil by outside characters located in the outlying cities in the country. For example, a local sheikh moved into the office of the director general at the Bayji Refinery, Iraq's largest refinery. There was a power vacuum in the Bayji area after the fall of Saddam and the sheikh felt that he could claim the refinery leadership job under this new democracy. Riyadh Ghassab showed the

DEPARTMENT OF DEFENSE
OFFICE OF RECONSTRUCTION
AND HUMANITARIAN ASSISTANCE
CFLCC-ORHA
APO AE 09304

May 3, 2003

To: Ministry of Oil

From: Oil Advisor, Office of Reconstruction and Humanitarian Assistance

Subject: Appointment of Oil Ministry Leadership – Interim Management Team

It is very important to stand up the Oil Ministry leadership in order to quickly return Iraq's oil infrastructure to efficient and safe production levels. This interim management team will be responsible for day-to-day management responsibility of the oil sector. Additional appointments will be made in the future to complete the team. The following appointments are effective immediately.

Chief Executive Officer – Thamir Al-Ghadhban
Director General, Economics & Finance Directorate – Dr. Radhwan Al-Saadi
Director General, Technical Directorate – Dr. Faleh Al-Khayat
Director General, Reservoir & Fields Development Directorate – Ali Hammadi
Director General, Department of National Manufacturing – Katham Al-Anbaki
Director General, Administration & Legal Department – Medhat Abbas Kaddori
Director General, Control & Internal Audit Directorate – Sharif Muhsin
South Oil Company Director General – Jabbar Al Li Aibi
State Company Oil Projects Director General – Hasam Al Deen Kashamola
Iraqi Drilling Company Director General – Zuhair Shakir
Petroleum Pipeline Company Director General – Kamil Al-Fatli
Oil Exploration Company Director General – Ibrahim Abbar Kareem
LPG Bottling Company Director General – Hussain Al Hadithi
North (Baiji) Refinery Company Director General – Riyadh Abd Ghasab
Central (Doura) Refinery Company Director General – Dathar Khashab
South (Basra) Refinery Company Director General – Thaer
North Gas Company Director General – Salah Aziz Karim
South Gas Company Director General – Najm Makhawar
Oil Products Distribution Company Director General – Samer Michael
Iraqi Tanker Company Director General – Falah Al-Asi
North Oil Company Deputy Director General – Adel Qazaz
State Oil Marketing Company Deputy Director General – Ali Rajab Hassan

Gary L. Vogler

Gary L. Vogler, Oil Advisor
Office of Reconstruction
And Humanitarian Assistance

Oil ministry appointment letter

appointment letter to the local sheikh and politely requested that the sheikh vacate his office.

Some on the original leadership team were replaced within a month, but many of them lasted for more than a year. Dathar Khashab, director general of the Doura Refinery, stayed in his position until his retirement at the end of 2010. Jabbar Al Li Aibi (also spelled Al Luaibi) remained as the head of the South Oil Company until 2009. He came out of retirement in 2016 to accept the position of oil minister. Dr. Radhwan Al Saadi remained in his position until 2007. Thaer stayed until 2008. Adel Qazaz remained the head of North Oil Company until his kidnapping in 2006. Sameer Michael was transferred from Distribution to the Technical Directorate in 2003 and remained in that position until 2009. Salah Aziz was transferred from North Gas to Product Pipeline in 2003 and remained in that position until 2009.

Thamir Ghadhban remained in the position of CEO until appointed as oil minister, which he served as from May 2004 until May 2005. He was elected as a member of the National Assembly in 2005 and then appointed to create and lead the Prime Minister's Advisory Commission (PMAC) in 2006. This commission of about seventy technically qualified senior Iraqis became the prime minister's principal advisory office for matters related to oil, electricity, water, education, agriculture, trade, planning, and finance. He continued to lead this influential group within Iraq's government until his retirement in 2015.

CHAPTER 8

The Calm before the Storm

June 2003

I agree. It is a no-brainer.
—Philip J. Carroll

The period from early May until mid-July 2003 was relatively calm from a security standpoint, enabling a lot to be accomplished within the oil sector. The Ministry of Oil was in capable Iraqi hands. However, we would learn that the infrastructure was falling apart. The sanctions regime of the previous decade had an enormous negative impact on the oil ministry's capability to sustain operations. Repair parts warehouses were empty and normal scheduled maintenance practices had disappeared. "Maintenance under fire" was the term used to describe how the ministry maintained equipment—run it until it catches fire.

In May, the ORHA oil team changed into the Coalition Provisional Authority (CPA) oil team. Task Force RIO had subject-matter experts visiting every oil site in the country, assessing looting and damages from the war. These assessments helped us eventually develop a detailed plan to repair and restore the oil sector to prewar levels.

CPA installed a new cell phone system in Baghdad during May, which was very helpful for communicating within the oil sector. CPA installed a trailer park in the back of the palace with electricity and running water for our lodging comfort. Times were looking good. Our biggest immediate problem was getting the Iraqi refineries operational so that they could produce gasoline, diesel, and other humanitarian fuels for the population.

Visiting the director general of Midlands Refinery Company, Dathar Al Kashab, became a high priority after appointing the leadership team on May 3. Al Kashab was educated in the United Kingdom and spent more than thirty years at the refinery following his short period in the army. The visit to the Doura Refinery was a high priority, since there were shortages of gasoline and diesel products throughout the country.

The Doura Refinery was located on the Doura industrial peninsula, across the Tigris River from downtown Baghdad. The flare from the refinery could be seen from just about anywhere along the Tigris River in Baghdad. The refinery had a capacity to process about 110,000 barrels of crude oil per day. It had three old crude oil units that had been initially installed back in the 1950s. It also had a few small package crude units that could handle ten thousand barrels per day and could be dismantled and erected on short notice.

The refinery used old technology that could only yield gasoline and diesel products of about 50 percent of the crude oil processed. The other 50 percent was primarily a heavy fuel oil that had limited use, the bottom of the barrel. There were no complex cracking units to improve the yields.

The poor yields of gasoline and diesel created significant operational and distribution problems. The 50 percent heavy fuel oil (HFO) created containment issues since there was very little demand for that product, resulting in the refinery storage tanks filling up quickly. There were plenty of customers to take the gasoline and diesel, but not enough customers to take the HFO. This supply-demand imbalance eventually restricted the volumes through the refinery and created a bottleneck for years.

The Doura Refinery was important to the ministry refinery sector as it was the first large refinery constructed in Iraq. Since it was located in Baghdad, it was the center of expertise within the refining sector. The other refineries at Bayji and Basra looked to Doura whenever they needed materials or expertise. It also had a strong leader, Mr. Dathar Al Kashab.

The flare of any refinery is an important piece of safety and environmental equipment. Refineries are designed to minimize the release of flammable gases into the atmosphere. These gases are captured within the system of pipes and sent to the flare for safe combustion. The Doura Refinery kept its flare burning, even when the refinery was shut down in preparation for the war. Why? Dathar made the decision to keep the flare burning, sending any hydrocarbon gas he had in storage to keep it burning. In his mind, the Doura flare represented much more to the citizens of Baghdad. It could be seen for miles around and was the only light in the city during the "shock and awe" bombing campaign. Dathar said it represented a "beacon of hope for the future."

Dathar impressed me with his energy, understanding of his refinery, and understanding of the technology he needed at Doura. He knew his refinery's strengths and weaknesses and he had the ability to articulate them very well. My recognition of his talents would be echoed over the next eight years by every American working in the oil sector, both civilian and military. Years later, his talents would be recognized by the private sector after his retirement. He has built refineries in Kurdistan for a private company since 2011.

Dathar also had a fabulous sense of humor. He shared many stories of the first Gulf War and how the Doura Refinery was severely damaged by the bombing campaign in that war. He spoke of how a rocket pierced through the top of one of his crude units without exploding. It just left a large hole. I could see the patch job from the refinery road. The crude unit had been repaired and placed back in operation in 1992.

As Dathar told the stories of the many bombs that had hit the refinery, he noticed that I started to become defensive of our pilots. From the pictures I saw, our US bombers did massive destruction to this refinery in 1991.

Always being the gracious host, he looked at me and said, "Gary, I can tell that I am making you uneasy by describing all the destruction caused by the American air force. Now that I think more, it was probably the French that did all this damage. From now on, I will always blame the French." True to his word, whenever I accompanied new visitors to Doura over the next eight years, Dathar would look at me with a smile and explain to the visitors how the French destroyed his refinery in 1991.

He spoke to me for several hours during my first visit. Late during the visit, I interrupted him to share a story.

"Dathar, I watched the initial part of the war at the CENTCOM headquarters in Qatar. I had access to satellite photos of this refinery during the first few days. I noticed that you disassembled your small package units of ten thousand barrels per day, and placed them on flatbed trucks just before hostilities started. I also noticed that you did not move them outside the refinery during the bombing campaign. I was happy that you did not move them. I feared that if some American F-16 pilot saw this large vessel moving down the highway on a flatbed he might mistake it for WMD and destroy it. Why did you keep those flatbeds inside the refinery?"

Dathar laughed, "Gary, I also spent time in the military as a young lieutenant early in my life. I have learned some military tactics. I knew that if you did not bomb the refinery during the first twenty-four hours, you would not bomb it at all. It was not on your target list. But please, Gary, the next time, could you phone to save me the trouble of dismantling my package units?" Everyone in the room laughed. This was the beginning of many visits to Doura and several encounters with a special person whom I still consider a dear friend today.

Phil Carroll arrived in Baghdad via a C-17 cargo plane from the United States along with several new members of the CPA during the second week of May. Phil was the former CEO of Shell US and the former chairman and CEO of Fluor Corporation. We greeted Phil in the rotunda and helped him with his bags to our small, very modest office space, with barely enough room to hold four people and four desks. Our request for more space was eventually approved and we moved to a larger office within a week.

Phil was taken to the CPA oil team's living quarters on the second floor. Clarke Turner had been able to convince some people that we needed bigger sleeping quarters, so we were assigned the room that had previously been the second floor palace library. The good news was that there was a small private room adjacent to the library where Phil could sleep. The bad news was that we still did not have running water or air conditioning. The palace swimming pool was outside our window, so there were plenty of mosquitoes!

Phil and I had a good discussion early the next morning. It was a meeting to get to know each other. His first comment to me was a bit of a surprise. Phil mentioned that if he saw anything approaching an oil grab, he would immediately resign and be on the next plane out of Baghdad. Recognizing the serious tone in his voice, I replied that I had not seen or heard anything close to an oil grab, but if I did, I would come to him and we would leave together.

It was good to hear this comment directly from Phil. It reinforced to me that this war was not about oil. It was not about enriching oil companies that might be friends of the administration, as the press was starting to report. I did not sign on to an adventure where American soldiers would sacrifice their lives so that some oil company could enrich its coffers. It was reassuring to hear that Phil and I were of the same mindset.

Phil further shared that he had taken the time during his stay in Washington to meet with senior officials within the Bush Administration to ensure that everyone understood his role while leading the oil sector in Iraq. As he said more than once, he felt that he had support from the top levels within this administration.

Phil originally agreed to serve in a position called chairman of the advisory board for the Ministry of Oil. In our worst-case scenarios while planning at the Pentagon, we had to be prepared for a case of massive destruction to the oil sector and a workforce that was not ready and willing to work with us. Fortunately, that scenario was avoided and it was time to adjust things accordingly, including Phil Carroll's public title. It was agreed that his official title would be director of oil policy and he would report directly to Ambassador Bremer. Phil continued to communicate with a small board located outside of Iraq, but he was beginning to think that such a board might not be needed.

After Phil arrived in Baghdad, the telephone calls with Pam Quanrud at the NSC converted into conference calls that included Thamir Ghadhban, Pam, the advisory board, and myself. The advisory board consisted of Phil Carroll as chairman, Fadhel Ali Othman as vice chairman, and Dr. Falih Aljibury.

Fadhel Ali Othman lived in Istanbul, Turkey. He worked in the Ministry of Oil for more than thirty years, escaping with his family to Turkey in the early 1990s. He had an excellent reputation within the ministry, similar to Thamir Ghadhban's.

Dr. Falih Aljibury lived in San Francisco. He was an Iraqi-American with a good reputation within our government for analyzing OPEC decisions and other Middle East energy issues. These conference calls proved quite helpful during the early months but were stopped in late July.

Phil was well known throughout the US oil industry. At sixty-five, he was a senior statesman in the industry, a former officer in the US Army, and a personal friend to former President George H. W. Bush. He was a person that everyone enjoyed working with, both Americans and Iraqis. There are many excellent stories that can be told about Phil Carroll's service to his country that summer. A great story about Phil is one that he has never publicly shared but should be told.

In May 2003, a new American army unit was assigned to provide security for the Ministry of Oil. Unlike previous units, this new unit from the 1st Armored Division would be assigned responsibility throughout the long hot summer of 2003 and into 2004. Phil recognized the importance of ensuring that this new unit started off on the right foot, developing a good relationship with the Iraqis within the Ministry of Oil. To foster this relationship, Phil instructed Thamir Ghadhban to have his people set aside a building located within the ministry compound for the new unit. That building was fitted with new air-conditioning units, a large-screen television set, a VHS player, a satellite receiver, stereo equipment, and weightlifting equipment.

The members of the unit quickly recognized how the Iraqis were working to provide these creature comforts and morale builders. The young captain commanding the unit and his NCOs became well known among the senior Iraqis, as well as those of us from CPA oil. It was a brilliant move on Phil's part. At the time, there was no legal way to pay for any of this material by either the CPA or the Ministry of Oil, at least not expeditiously. Phil took out his checkbook and wrote a personal check for $17,000 to pay for everything. He would later tell me that it was one of the best investments that he had ever made.

Phil's time commitment was for six months, until the end of September 2003. Because of his age, the Iraqi heat became intolerable, so he spent most of July and part of August outside the country, in either the United Kingdom or back in the United States. We maintained daily contact over the phone. He was the face of American oil in the eyes of the press during the entire summer, relieving me of much of that responsibility. There were many instances that I found his comments to the press quite amusing.

Sitting in our office one afternoon, a member of the press started asking Phil questions. The young lady mentioned that there were rumors in the US press that senior members of the Bush Administration were saying that it was a "no-brainer" for the United States to privatize the Iraqi oil sector. The reporter asked for Phil's comment. He immediately responded, "I agree, it is a no-brainer." A long pause occurred before his next words were uttered. My jaw dropped when I heard those first words come from his mouth. Then, he finished his point: "Anyone thinking like that in the US must have no brains."

In another interview by the press, Phil was asked about forecasting where crude prices were headed over the ensuing six-month period. His response was classic.

"Whenever I believe that I can accurately forecast the price of crude oil, I immediately go straight home, take two aspirin, and go to bed," Phil replied.

Phil did not have an email account. CPA tried to assign him one, but he ignored it. He felt more comfortable communicating face to face, on the phone, or with letters. One exception he made was to former President George H. W. Bush. He got Clarke Turner to be his messenger. Clarke wrote emails to Barbara Bush, who shared with former President Bush. Many of those emails were personal commentary about how things were going and sending pictures of Phil's time in Baghdad.

Phil's first trip to the Ministry of Oil took place a few days after his arrival in Baghdad. By this time, we had started to use the large SUVs that ORHA had purchased in Kuwait for our transportation needs. They did not have any armor or special glass. That would be required later in the summer. We could travel in those vehicles as long as there was one person carrying a weapon and there were two vehicles in a convoy.

As we passed by the large Galani gas station along Route Irish about one mile from the ministry, we noticed that there were long lines awaiting gasoline. Gasoline and diesel shortages were throughout the country and represented one of ORHA/CPA's biggest early challenges. Phil pointed to the long lines and said, "Gary, that line is one of our key performance indicators. When that line is long we need to work harder. We are not performing well enough."

Phil was correct. Whenever there were shortages of gasoline in Baghdad, the Galani gas station was one of the first to experience long lines, lines several kilometers long. The shortages of gasoline and other petroleum products contributed to a lack of confidence in CPA and instability within Iraq. It literally translated into casualties of US soldiers.

This theory was validated in August 2003. The commander of the 101st Airborne Division, Major General Dave Petraeus, was asked about how the shortages affected his unit. General Petraeus shared with me the story of how one of his infantry platoons was trying to control the crowds awaiting distri-

Galani gas station located along Route Irish—the main route between the Green Zone and the oil ministry, May 2003. Photo by author.

bution of scarce petroleum products in downtown Mosul. An Iraqi walked up behind a young American soldier, put a pistol between his helmet and flak jacket, and blew his head off.

Similar stories from other military officers were told, but this story from General Petraeus was described in such detail to me that it remains chiseled in my brain. I was reminded of this horrible story every time I passed that gas station in Baghdad and saw long queues with US soldiers controlling the crowds.

Phil went to Thamir Ghadhban's office on the fourth floor of the ministry headquarters. Phil and Thamir had previously spoken by telephone, but this was their first encounter. Thamir was always a gracious host. The two of them talked while I excused myself for another meeting. When I returned a couple of hours later, they were still busy talking.

Phil recognized that it was time to return to the palace, but he advised me to take the rest of the team and return without him. He wanted to continue meeting for a couple more hours. When Phil showed up at the palace after 5 PM, he told us that Thamir had walked him out to Route Irish in front of the ministry and hailed a rusty, orange and white taxi. Thamir advised the driver to deliver Phil to the palace. This would have been much too dangerous for any American in a few short weeks. Things were still relatively calm.

Addressing the shortages of gasoline, diesel, and cooking fuel was our number-one priority in May. A meeting was held on Friday morning, May 16, at the Ministry of Oil cultural center in downtown Baghdad. The meeting was cochaired by Thamir Ghadhban and Phil Carroll. It became the first of several Friday meetings, until we worked our way through this first fuel crisis.

Task Force RIO and KBR played an important role in supplying product from outside of Iraq. Product was trucked in from Turkey and Kuwait. KBR not only sourced the product but also played a key role in preparing the depots to accept trucked material.

Iraq's distribution depots had previously received their products via pipeline. The pumps were designed for loading trucks for distribution to the gas stations. Now they needed to accept trucks to offload product into their storage tanks. Each depot had to be fitted with the proper pumps, valves, hoses, and generators to facilitate this undertaking. Two senior engineers from KBR traveled to some high-risk locations with little security to accomplish this high-priority job. Their names were John Bauer and Daryl Holland. They received top priority from Ghadhban and his leadership team to complete their work.

As the convoys of trucks loaded and staged at the Kuwaiti border, the security of the convoys became an increasing priority. Coordination between Task Force RIO, CPA oil, and the logistics element within CFLCC had to be addressed. Major General Claude "Chris" Christianson was the senior logistician under General McKiernan in CFLCC. Christianson expressed concerns about the capacity of the Iraqis to adequately receive and distribute the large convoys. Two to four hundred tanker trucks a day were scheduled into Baghdad. He sent us one of his best planners, Major Rebecca Freeze.

Rebecca's first job was to help us quickly get Christianson comfortable so that he would support the RIO import operation. We held a meeting at the oil ministry headquarters during the third week of May, chaired by Thamir Ghadhban. General Christianson sat to Ghadhban's right and observed Ghadhban while he directed several of his Iraqi subordinates. He tasked them on how to receive the trucks. Late in the meeting, it became obvious that the Iraqis were ready. Ghadhban looked to General Christianson to see if he had any questions. Chris smiled and said he would contact his people to let the trucks roll.

This was the first fuel crisis we had to overcome. Although the massive imports helped to decrease shortages in the Baghdad area, many of the outlying areas continued to experience shortages of varying degrees for several more weeks until the refineries at Doura and Bayji became operational again.

Major General Claude "Chris" Christianson, Gary Vogler, and Thamir Ghadhban at oil ministry logistics meeting in May 2003.

As new automobiles and trucks were imported into the country to satisfy the demand that had been pent up for the previous twelve years, and as the use of gasoline generators increased during the summer months, we recognized that the May crisis would not be our last. Major Freeze stayed for a couple of months and played an important role. USAR Colonel Gary Major augmented her function in June and he took over responsibilities later that summer. Rebecca moved to another logistics assignment in July at Balad Air Base, north of Baghdad.

The State Oil Marketing Organization (SOMO) was the marketing arm of the Ministry of Oil. They were responsible for selling all oil exports. SOMO had been selling oil for decades to the international markets and had a staff of very capable marketers with advanced degrees from universities in Europe and North America. At about 150 employees, it had the highest per capita level of education within the oil ministry.

The SOMO compound was located about a mile from the oil ministry headquarters and consisted of a three-story office building and enough housing for most of the SOMO employees. It had a small wall around the perimeter and one gate for entering the compound.

A few days after the fall of Saddam, a couple hundred looters entered the compound and looted the office building. The offices containing the important sales records were set ablaze. The stench from the smoke was still strong during the second week of May 2003.

The director general of SOMO was one of the first to be replaced from the appointments made on May 3. He had connections to the Baath Party and he was not highly thought of by the rank and file within the SOMO organization because of the business practices that were allowed while he was the deputy director general under Saddam. Mohammed Al Jibouri became director general in May after the SOMO employees used their newfound democratic freedoms to vote for a new leader. Thamir Ghadhban explained this new selection process of employees voting for their director. Thamir told me not to worry. He knew who was leading in the early vote count and it was the person he would have selected to lead SOMO.

Phil Carroll and I made our first visit to Mohammed Al Jibouri in late May to discuss policies of marketing crude oil under CPA. Every time we brought up the subject of crude oil sales with Thamir Ghadhban he would ask us to defer any discussion until after we had met and talked with Mohammed. So this first meeting was one that both Phil and I were looking forward to.

After the initial introductions, sweet tea (Chai), and pleasantries, Mohammed moved into a discussion of how he wanted to market crude oil during his time as director general. Mohammed was educated in the United Kingdom and had been the senior oil sales manager within SOMO for many years. He was well known among the large international oil companies. He quickly shared a plan with us that was exactly in line with the way Phil and I had discussed before the meeting.

The plan was to follow a marketing strategy very similar to the way that Aramco, Saudi Arabia's national oil company, had been marketing their crude oil over the previous fifteen years. The crude oil would be sold under contract only to large end users, refinery owners. Iraq's crude oil would be priced based on market pricing in three large enclaves: East of Suez, Northwest Europe, and the US Gulf Coast.

This was different from the way Iraq had marketed its oil under the UN Oil-for-Food program. The UN program of selling to non–end users, traders and brokers, allowed for the high level of corruption that prevailed under Saddam and the Oil-for-Food program.

Selling the crude oil only to end-user refineries also had the added benefit of avoiding traders who could easily game the pricing system and intentionally advise that their shipment was going to a low-price enclave such as the United States but actually ship the cargo to a high-price enclave such as

Singapore. A price differential of a dollar per barrel on a two-million-barrel vessel was substantial.

This gaming by traders was well known in the Middle East crude oil marketing business and contributed to corrupt practices. When traders were caught gaming the system, their trading firms were blacklisted. These companies would close their business, move locations, change their name, and reopen a few months later with the same people and using the same business practices. Large, multibillion-dollar refineries, on the other hand, do not have the ability to move to new locations.

Mohammed was well aware of how some well-known unscrupulous international players had previously contributed to Iraq's corruption problems under Saddam. He was not going to allow those practices under his watch.

Phil and I drove away from our first SOMO meeting with a high degree of confidence that SOMO was in good hands. We had similar meetings with Mohammed every week until the end of June. Each time we pulled away in our vehicle, Phil would comment about how lucky he felt that someone of Mohammed's reputation and integrity was leading SOMO during our watch.

Once Mohammed took over, our biggest challenge with SOMO during the early days was getting adequate communications. Fax machines were used to sell the first shipments of crude oil and we wanted those ships moving in June. We needed a satellite system with much more capacity than the Thuraya phones that most of us carried. A small Inmarsat commercial system was recommended. It became a challenge to locate and procure an Inmarsat satellite system capable of commercial fax through the US government procurement system.

Two satellite systems were located in Dubai that the supplier could ship immediately for about $20,000. That proved to be the easy part. Getting priority support from CPA procurement was a bigger challenge. Ambassador Pat Kennedy was Bremer's senior administration director. Pat's "can-do" attitude impressed me and I went to Pat whenever we needed help with our ever-expanding bureaucracy. Kennedy delivered.

The satellite system arrived, but there were no instructions. Nothing was easy in Iraq. It was the same system being used by the international press to transmit their stories back to their home stations. A reporter was using one in front of the palace to transmit his story, so I asked if he would give me a quick lesson on how to operate it.

The system was successfully installed at the SOMO headquarters in time to receive bids from the international oil companies competing for Iraq's first post-Saddam oil shipments. Tupras of Turkey won the first one million barrels, lifting out of the north at Ceyhan on June 23. Chevron was the winner for the south and lifted their two-million-barrel cargo from Al Basra oil terminal (ABOT) on June 30.

Brigadier General Bob Crear and his Task Force RIO team were busy during May and early June conducting assessments of all oil installations throughout Iraq. Our subject-matter experts from KBR, along with US Army officers on the RIO team, inspected every facility that had some type of damage.

Crear related a story of one site that could not be inspected at first attempt. Flying into the "Saddam GOSP" located east of the Bayji Refinery in a Blackhawk helicopter, the pilots noticed that there were hundreds of looters at the site. The helicopter pilots refused to land the birds for security reasons. They hovered a few hundred feet above the crowd; Crear commented that it looked like a scene of several hundred ants picking at a chicken bone on the ground. The looters were picking the place clean; anything not bolted down was hauled away. The looters were loading trucks full of material from the site.

Fortunately, most of the sites did not pose this type of security problem. The assessment teams entered the facilities, reviewed the damage, took pictures, and returned their results to a central location for estimators to analyze and estimate the full cost of repair.

While the RIO assessment teams were quickly moving throughout the country to assess damage, RIO assigned another team in Baghdad to prepare for a conference in mid-June. The conference participants included representatives from each of the oil-ministry-operating companies throughout the country, managers at the Ministry of Oil and Task Force RIO/KBR, and a few of us from CPA oil. The conference was broken down into groups with responsibility for North Oil Company, South Oil Company, North Gas Company, South Gas Company, refineries, Product Pipeline Company, and the State Company for Oil Projects (SCOP). Each group reviewed the damage and the estimates calculated by the RIO team.

At the end of the conference, there were 224 projects identified that needed to be successfully completed for the US government to meet its commitment of repairing and restoring the oil sector to prewar levels. The total cost of these restoration projects was about $1 billion; the looting that followed the war caused 80 percent of the damage.

Phil Carroll reviewed the final list of projects. That was when Phil discussed the repair of the K-3 pump station in Anbar Province. He told me that he had his orders from the highest levels at the Pentagon. This pump station would not be repaired. The $50 million assigned to repair that pump station was offered to Thamir Ghadhban for any projects he wanted completed for the same amount of money.

Phil Carroll and Thamir Ghadhban approved and signed an agreement that the US commitment to Iraq would be honored once these projects were completed by Task Force RIO. Our plan was in place and it appeared to be an excellent plan. Unfortunately, execution of that plan became extremely difficult when security deteriorated, funding was cut, and priorities changed. The US government completed only a small percentage of the 224 projects.

Also in mid-June, the first formal CPA oil report was submitted to Ambassador Bremer. It was an assessment by Phil Carroll of the current oil sector status and a look at the next six months. It was a six-page report covering several details of current issues. The summary was positive and stated the following: "The ministry should be ready by 1 January 2004 for immediate absorption into a permanent government of Iraq, whenever that government is formed. Strong and competent leadership is in place and most major facilities are running although many are not at full capacity."[1]

Phil's forecast of oil exports was an increase from an average of 750,000 barrels per day in late June to two million barrels a day in December 2003. Estimated revenues from oil exports would be about $6 billion in 2003.

Phil forecasted the personnel resources required to support CPA in the oil sector for the balance of the year. Our resources located in Basra and Kirkuk moved to Baghdad or went home after their tour. The following was the Baghdad office staffing.[2]

June 2003 Staffing
P. J. Carroll
G. L. Vogler
C. Turner (DOE)
J. Kjar (Australia)
D. Callahan (DOE)

July 1–September 30 Staffing
P. J. Carroll (part time)
G. L. Vogler—senior advisor
G. Holcomb—Task Force RIO liaison (DOE)
Josh Mikesell—advisory board liaison & administration (DOS)
Public Info Officer—to be identified
Iraqi Support Staff—to be identified

While all of the other CPA ministry advisory staffs submitted plans to Ambassador Bremer for sizable increases to their headcount, CPA oil submitted a plan to downsize. Our staffing plan was substantially different from the others. Most Iraqi ministries were in bad shape. Their institutions were either broken or close to being broken. The Ministry of Oil was in fine shape at that time.

Phil's resource plan included a leave schedule sending me home during the last week in June and returning by July 1, allowing Phil to take an extended absence from Iraq during the hot months.

Little did we realize at the time that we were experiencing a calm before the storm. Things would change dramatically in about a month.

CHAPTER 9

The Long Hot Summer of 2003

No WMD, so why are we here?
—Ambassador Robin Raphel

While I was landing in Kuwait on July 1 after my short visit home, Phil Carroll and Clarke Turner were departing on a C-130 air force cargo plane back to the United States through Europe. Clarke was leaving permanently and Phil was traveling to London for a break from the war.

Earlier in the day and before their flight, they witnessed the loading of the first shipment of crude oil out of the south from the offshore export terminal. As they returned to Kuwait and drove their car out to the US air base, Phil was in a festive mood. They were both leaving Iraq and they had just witnessed the first two-million-barrel very large crude carrier (VLCC) depart the terminal. They were celebrating their successes.[1]

Phil and Clarke loaded their bags on the C-130 and sat in the nylon jump seats, continuing their cheerful conversation. After two months in a war zone and several early accomplishments in the oil sector, both were feeling celebratory. Then, a team of soldiers carried a casket onto the plane. It had an American flag over the top and it was secured to the aircraft flooring in front of the two of them. It was one of the thirty US fatalities during the month of June. Not another word was spoken during the next five hours of the flight out of respect for this soldier who had made the ultimate sacrifice.[2]

The next ten weeks in Iraq would become the most physically and mentally challenging in my life. Temperatures in Baghdad averaged more than 110 degrees during the late-summer period. Summer in Baghdad and most of Iraq was hot, very hot. Labor unrest in the oil sector, de-Baathification, a refuted justification for war, wrongful arrests of oil ministry personnel, attacks on oil pipelines, product shortages, Basra riots, and distractions from the Pentagon were some of those challenges.

Labor Unrest

Before catching a flight to Baghdad from Kuwait City on July 2, I checked in with General Bob Crear. He and his engineers were still headquartered in Kuwait City. Bob's report on the reconstruction effort was upbeat and positive. His engineers were moving quickly to address the 224 projects that had been identified in mid-June.

The one issue that Bob said was on everyone's mind in the oil sector was payroll. He asked me to address it with CPA upon my return to Baghdad. He said that it was an issue that was festering within South Oil Company and the other oil ministry companies. Bob reported that his people were hearing some pretty hostile comments made about CPA concerning the payroll issue.

Anyone who has worked in industry understands the importance of properly paying employees. As a young engineer at Mobil Oil in the early 1980s, I quickly learned the importance of paying employees properly and on time. I knew this issue had to be fixed quickly. Upon my return to Baghdad, a meeting was scheduled with Dr. Radhwan Al Saadi, the oil ministry's director of finance, to ensure that he was ready to change from the temporary payroll system put in place under ORHA.

Colonel Bob Schroeder had taken on the responsibility in June of reviewing the old payroll system to understand how it worked and gain confidence that we should be transitioning the oil ministry back to their old system. Up until that point, ORHA and CPA had been paying all government employees under a temporary four-tier system. Any government employee was paid one of four amounts, depending on their grade level. This procedure was a stopgap system established by the ORHA treasury team to put money into the hands of government employees until we became comfortable with the old Iraqi government payroll system.

The oil ministry was ready to transition during the month of July. Bob had reviewed the payroll system and pay records for the entire ministry. He explained their system to me. It was a sophisticated "pay-for-performance" system that was as good as or better than anything that I had seen during my time in the oil industry. It had over a dozen salary levels, similar to most Western companies. It had adopted a bonus system that paid year-end bonuses to employees. These bonuses were performance-based and significant. The bonus pool was based on the profitability of the operating company. Shares in the bonus pool were awarded to management and low-level employees based on their performance during the year. Employees could achieve bonuses that were several times higher than their base salary. The one change that we would make was to eliminate the Baath Party bonus, a minimum award made to employees who were members of the Baath Party.

Bob and I met with Dr. Radhwan concerning timing of the payroll change, and everything was ready to go, at least that was our impression. Our challenge was to get approval from the CPA budget director, retired Admiral David Oliver.

My meeting with Oliver did not go well. He insisted that he was going to install a new civil service payroll system similar to that of the US government system. He recognized that the oil ministry was ready to implement their old system, but he refused to approve it. He told me there would be little incentive to accept his new civil service system if the oil ministry was allowed to use the old Iraqi payroll system. The hubris of his comments defied logic. I would experience several cruel incidences of American arrogance during my time in Iraq. This would rank as one of the worst. We warned Oliver that any delay to a proper payroll system would have dire consequences for the oil sector; labor strikes would happen and soon. He either did not believe us or he had an ideologue agenda.

Phil Carroll was at his home in London at the time. During our daily phone call, I mentioned to him that I was resigning. If Bremer's staff refused to listen to us, then I was prepared to resign and go to the press with my reasons. Phil convinced me to stay. He would handle Oliver. Phil got on a plane and returned to Baghdad at the end of July for a short visit to discuss the payroll issue. It was too late.

The first labor strike occurred in South Oil Company. Oil exports were stopped for a few days. Oil revenues suffered by about $50 million because of the strike. The payroll issue created an opportunity for the weak labor union in South Oil Company to gain power. The union declared a strike. It was the first strike in the history of the Ministry of Oil. Oliver's decision was a significant blunder.

Several other companies within the oil ministry went on strike as well. In early August, the Bayji Refinery workers told their director general to leave the refinery and threatened to kill him if he returned. Bayji was the largest refinery in the country, supplying 60 percent of the domestic gasoline and diesel products.

Ghadhban phoned one evening and advised me that we needed to go to Bayji early the next morning. The employees had revolted and were running the refinery. We needed to reestablish competent authority.

My first thought was to request help from Brigadier General Frank Helmick, deputy commanding general of the 101st Airborne Division. The 101st had a great "can-do" culture. Frank had hosted me during a visit to Mosul in early July to meet his boss, Major General David Petraeus. Both officers had impressed me with their understanding of the importance of providing essential services to their area of operation in Mosul. Shortages of gasoline, diesel, and cooking fuel contributed to instability and attacks on US troops.

Petraeus had assigned Helmick responsibility for oil and electricity in their area of operation. Frank and the local oil ministry distribution manager, Mohammed Zibari, had become inseparable. They came to the oil ministry many times to get support for oil initiatives in their area. Thamir Ghadhban remarked several times during the summer of 2003, "If the US had more generals like Frank Helmick, the oil sector and Iraq would be in much better condition."

Helmick arrived in Baghdad at 0630 hours with two Blackhawk helicopters to take three Iraqis and me to Bayji. Upon arrival at the refinery, we quickly conducted a preliminary meeting before seeing the employees who were leading the revolt. We entered the large meeting room on the ground floor of the refinery headquarters. It was a room that would normally hold about forty people comfortably. There were over a hundred waiting for us, standing room only. Thamir Ghadhban, Abdul Jabbar Al Waggaa, Dheyaa Al Bakkaa, and I entered the meeting room and sat at the head of the long table.

Ghadhban brought Jabbar Al Waggaa because of his leadership skills, but more important, Jabbar was the son of the former grand sheikh of the very large Al Jibouri tribe. His father had passed a few years before and Al Jibouris from all over the Middle East had come to Mosul to pay their respects to their late sheikh. This tribe had more than four million members, both Sunni and Shia. Thamir felt that he needed someone of Al Waggaa's stature to help with this critical problem in the middle of Sunni territory. Thamir would appoint Al Waggaa in another year as the deputy minister for upstream. He would be kidnapped in the summer of 2007 by the Shia death squads and released only after the Shia members of the Al Jibouri tribe pressured the militant Shia cleric Muqtada Al Sadr.

Dheyaa Al Bakkaa was the office manager for Ghadhban and a trusted advisor. Dheyaa's career was primarily in SOMO, where he demonstrated good communication skills and a great ability to solve problems. Ghadhban would appoint Dheyaa as the director general of SOMO in mid-2004.

The people in the meeting room looked hostile. My personal safety was at risk. Many had long beards and looked like followers of Wahhabism, a fundamentalist sect within Islam. There was a look of rage on all of these workers that I had never seen before on an employee's face. I was convinced that the wrong words would ignite a riot from this group. I feared for my life.

Ghadhban started speaking in such a way that immediately caught their interest. Dheyaa was sitting next to me and offered to translate for me. A translation was not necessary; I understood body language quite well. After ten minutes of talking, the look of rage turned into something less hostile. After thirty minutes, many workers began to smile. My anxiety level decreased. All were focused on every word said by Thamir. After an hour, they were all smiling and laughing. We broke up the meeting and many of the employees approached Thamir for hugs and friendly comments.

We left the headquarters and walked toward our vehicles to drive us to the helipad. I looked at Thamir and commented, "I don't know what you said in there, but it saved the day and my life. Thank you." Thamir looked at me, smiled, and jokingly replied, "Gary, I just blamed everything on you!" We both laughed and continued our walk.

Payroll issues continued to hurt oil operations until Ambassador Bremer agreed to grant a 10 percent pay increase to all employees in the oil and electricity ministries and committed to implement a new system in September. It was a temporary fix that seemed to work. Two payroll experts arrived in Baghdad in early September to implement this new system. One was a private sector payroll consultant and the other was Mrs. David Oliver, Admiral Oliver's wife. She had worked government payroll in the United States.

De-Baathification

The CPA De-Baathification Policy was announced in mid-May, but it took a couple of months to execute the order in the oil ministry. Phil Carroll and Ghadhban had discussed the impact to the oil sector extensively during June. Bremer had agreed to allow several exceptions for the oil ministry after Phil convinced him of the importance of keeping several key managers that were Baath Party members.

July was the month when the Baathists were dismissed. Upon arriving for my daily visit to Ghadhban's office in early July, a female employee walked out of Thamir's office just as I entered the outer reception area. The lady was carrying a baby in her arms, a cute little boy of about twelve months. The young mother was in tears: "How am I going to provide for my baby?" She was looking at me when she asked the question in broken English. "I have no husband and now I have no job. You tell me. What should I do?"

It was a bad scene. I did not know what to say. The baby was so cute and watched me with his brown eyes. His mother was having an emotional breakdown. The only thing that came to my mind was the money I had in my wallet. It was a couple hundred dollars. I pulled it out of my wallet and said, "This is all I have." She grabbed it from my hand and walked away.

As I walked into Ghadhban's office, he was visibly irritated. "That is the impact of your De-Baathification Policy," he said. "She lost her husband and now her job. There is obviously no compassion in such a policy."

There was nothing I could say to Thamir. It was not my policy and he knew it. Thamir was a strong nationalist, a leader, but even more so, he was passionate about the suffering of Iraqis and always showed compassion to those that were maltreated. He needed to vent his frustration and I was in the room.

My opportunity to voice opposition to the policy came later that evening during a meeting to discuss the subject. Meghan O'Sullivan was appointed by Bremer to execute this very unpopular policy. It was smart of Bremer to appoint someone as nice and sharp as Meghan. No person could be cross with Meghan. At least, I could not.

We met outside Bremer's office in the rotunda and reviewed the list of about twenty names and examined why each candidate needed a waiver to the policy. Meghan recognized the heartlessness of the policy. "Why is this necessary?" I asked. "Garner's policy seemed to be working well. Why such draconian measures?"

Meghan looked at me and shared her thoughts. She tried to rationalize the CPA policy. She mentioned that it might facilitate a reconciliation process later on. The Iraqis needed to demonstrate a willingness to compromise and forgive the technocrats who joined the party, especially those not part of Saddam's circle of thugs.

The reconciliation she described never happened. Most of the ministry employees who received waivers to the De-Baathification Policy from Bremer were dismissed before the end of 2003 by the new ministry leadership that arrived in September. The rest were dismissed within the following couple of years.

Justification for War Refuted

A meeting was held every morning at 0730 in the palace's south wing conference room for the senior ministry advisors. Ambassador Robin Raphel chaired these daily meetings and they quickly became one of our primary tools for communicating between the ministries.

Ambassador Raphel was a thirty-year career Foreign Service officer who had spent most of her career in South Asian countries like Pakistan and India. Robin had been vice president of the National Defense University (NDU) in DC just before joining the ORHA team earlier in the year. She did a nice job of keeping the daily meetings effective, meaningful, and appropriate. One such meeting in early July discussed the existence of WMD.

Two US Army colonels presented the results of their investigation of the mobile laboratories located in the northern part of the country. Secretary Powell's testimony before the United Nations earlier in 2003 highlighted that these laboratories were rolling labs used to manufacture biological agents, weapons of mass destruction. It was our justification for going to war.

Some Iraqis within the Ministry of Trade had told Ambassador Raphel that the so-called WMD labs were owned by the Ministry of Trade and were not used for WMD. She requested an investigation by the US military to look at this issue.

The report from the two officers took about ten minutes and identified the rolling labs as nothing more than quality-control labs operated by the Ministry of Trade with the purpose of testing food being imported under the Oil-for-Food program. Just before they concluded their final comments, Ambassador Raphel interrupted, "So there is no WMD, why are we here?" There was complete silence in the room.

"Good God," I thought. "What have we done?" We have destroyed a country and killed thousands, hundreds of US military have died or were severely injured, and why? Someone in my government did not do his or her job. Someone needs to be held accountable for this massive screwup. The entire country has been misled. No American should ever be placed in this position again!

My thoughts shifted to what I should do personally. What should we all do now that our justification for this war was mistaken? Secretary Powell's words echoed in my head, "If you break it, you own it." He was right. At this point, my country had a moral obligation, a duty to fix what we had destroyed. It was not the reputation of the Bush Administration that was at stake. It was the reputation of my country and what we stand for.

Wrongful Arrests of Oil Ministry Personnel

An arrest of oil ministry personnel by the coalition forces was not uncommon during the first couple of months. The default position of the military was to detain the people and let the military intelligence personnel sort out who should be released.

One specific incident that created a real problem was the detention of Kamel Fatley in early July. He was the products pipeline director general. Kamel and his brothers were older gentlemen, all in their sixties or seventies. They were summoned to their hometown of Diwaniyah for a tribal grievance from a small tribe in the area. After a couple days of tribal negotiations, it became evident that the smaller tribe was losing their case, so they sought help from the newest and biggest tribe in the area, the US Marine Corps.

The marines forced their way into the Fatley family home in Diwaniyah, where Kamel, his older brothers, and their sons were staying. The marines arrested all nine adult males. The opposing tribe claimed that the house was full of Baathists conspiring to attack the US forces. The marines believed the story, detained all nine, and labeled them as high-value targets, to be sorted out by the intelligence personnel at a future date.

Kamel's products pipeline job was extremely important because he was the key person in delivering gasoline and diesel products to the depots for distribution throughout the country. He had attended all the early meetings to help us fix the fuel shortage crisis of May. He was extremely important for the job of getting products throughout the country. Now, he was sitting in a detention facility close to Diwaniyah.

Thamir Ghadhban was incensed when this happened and became more irate the longer it took to get Kamel released. Our first order of business every time I visited Thamir or whenever I phoned him was to explain what I was doing to get Kamel released.

That was when I first met Brigadier General Janis Karpinski. She was responsible for all the prisons and later became infamous for the Abu Ghraib prisoner torture scandal. She was responsible for the detention camp in Diwaniyah and she reported to Lieutenant General Rick Sanchez. Karpinski was of little help. She told me that she did not have the authority to release any high-value prisoners. Her guidance was that Kamel would have to stay in prison until the intelligence folks got around to his case and that might take months.

Such details reported back to Thamir only annoyed him more. The loss of Kamel was critical for operations. Thamir had to reassign one of his strongest leaders from Kirkuk, Salah Aziz, to lead the pipeline company.

It became a bigger priority for Thamir when he heard that the marines lost Kamel's blood pressure medication and that Kamel had a mild stroke while in captivity. He had to be hospitalized for a week. It was now a humanitarian issue. The importance to Ghadhban increased exponentially.

My next move was to approach the chief of staff for General Sanchez and bypass Karpinski. I called on Brigadier General Dan Hahn and several of his senior colonels. Unfortunately, Dan and his team refused to help and only placed roadblocks in my path.

Ghadhban was becoming more irritated. I arranged a meeting between Thamir and Bremer to make his case for Kamel's release. That worked. Bremer directed Hahn to either produce evidence to support further detention of Kamel or immediately release him. General Hahn did not take the decision well. The look in Hahn's eyes as he left Bremer's office made me suspicious that he was not finished with his roadblocks.

Thamir told me that he and Dathar Al Kashab would make the two-hour trip to the detention camp to get Kamel. He knew that I could not travel with them because of the CPA security protocols, but I insisted on joining them. I suspected that this was not going to be an easy task.

Upon arrival at the detention facility, we were informed that the military orders from General Hahn stated that Kamel could be released only to a US military officer. I smiled when the New York National Guard sergeant told me in his strong Bronx accent that he was under strict orders and could not release Kamel to us. He could be released only to a US military officer. He showed me the written order signed by General Hahn. My instincts had been confirmed. I pulled out my wallet and produced my army reserve officer ID card, smiled at the NCO, and stated, "Sergeant, this is an easy one for you. As a US military officer, I am commanding you to release Mr. Fatley."

Kamel was brought out to us from the detention facility. He did not look good. He was weak and looked like he had lost several pounds since I had last seen him. He had been in detention for a month. It was still a joyous reunion. Dathar and Kamel had been close friends for decades; everyone was a bit misty-eyed.

Kamel approached me a few weeks later about getting compensated because the marines had confiscated his SUV when they arrested him and they could not locate it upon releasing him. We got Kamel a nice cash settlement through the military claims office for his vehicle and his pain and suffering. The JAG captain who gave Kamel his cash settlement said, "Mr. Fatley, on behalf of the US military, I apologize for what happened to you and hope that this cash payment helps to compensate you for our mistake." The captain did not have to be apologetic, but he demonstrated some real class when he did. Kamel appreciated it more than the money he received. It was a victory for the army claims office and that specific JAG captain. Kamel retired after that incident. His health was never quite the same.

Basra Riots

Baghdad was hot during the summer of 2003, but Basra was even hotter. Basra lost significant power-generation capacity when a fire broke out and severely damaged one of the larger power plants at Al Hartha. The power outages shut down the Basra Refinery, cutting off the primary supplies of gasoline and diesel products to the people of Basra.

Iraqis in Basra had become accustomed to power outages. They seldom received full power from the national grid before 2003. However, most of the citizens had their own generators for backup when power was not available from the grid. They needed gasoline or diesel to run these generators, but there were no supplies of these fuels with the refinery shut down. The local population lost their patience; demonstrations and riots followed.

Ambassador Bremer dispatched Major General Carl Strock and a team to Basra to fix the power and fuel problems. I was the oil representative on his team. We landed at the Basra airport on a day when the temperatures eclipsed 125 degrees. It was like an oven on the runway at the Basra airport.

We met with the British military unit headquartered at the airport after we checked in and were assigned a place to sleep. Major General Graeme Lamb was the senior commander of UK forces at the time. The airport was west of Basra, far enough away from the city and the riots. The British provided security and all of our logistical support during the visit.

Our team left early in the morning for various locations to assess the essential service problems causing the riots. I went to the refinery. It was located close to the airport.

The Basra Refinery processed about 160,000 barrels per day of crude oil. It was surrounded by desert and was several miles from any populated area. As our convoy approached the refinery entrance, I was surprised to see a few dozen demonstrators close to the front gate. We were quickly moved through the crowd and inside the complex.

There was a complete power outage, so the only noise I heard when I got out of the vehicle was the sound of the wind. Several employees were at work, but they were just waiting for the electric power to return from the grid. The refinery manager was waiting for me in front of the administration building.

"Mr. Gary, I am so happy to see you again. It is bad here. The people want to kill me," Thaer stated.

Looking at Thaer, I realized that he was terrified. "Let's go to your office so we can talk," I said. Thaer escorted me into the building and up to his second-floor office. His office was very warm and stuffy. Without electricity, there were no fans, air conditioning, or lights. It was a meeting that I will never forget.

We sat at a small table in front of his desk, just the two of us. He put a box of tissues in front of us and two bottles of water. We were perspiring so profusely that we grabbed a new tissue every thirty seconds to wipe our foreheads.

"Mr. Gary, those people at the gate want to kill me. They have threatened my family and me. I can't do this anymore. I quit, Mr. Gary." Thaer was quite scared.

Thaer was about ten years younger than me, in his early forties. We had appointed him in May, the youngest director general that we appointed. He was now faced with a very difficult leadership problem. The local population was blaming him for the shortages of gasoline and diesel.

I looked him in the eye. "Thaer, you cannot quit. I will not allow you to quit. You cannot quit until I quit, and I am not ready to quit. Now let's talk about how we get this fixed." It worked. He changed his tone and started talking about the operation of the refinery.

"The units are working fine. Our only problem is power and the electricity commission has not been able to provide any power to us. We need power. That's all we need." Thaer was speaking with confidence. "The operators were doing a great job even though we kept having power outages every few days. This last outage has lasted several days and I am hearing that we won't get power anytime soon. Mr. Gary, you need to get me power."

We continued to talk for another half hour or so and emptied the tissue box. Thaer walked me back out to the front of the building and I guaranteed him that we would get him his power, but we agreed to talk daily via phone until that happened.

Our meetings that night at the British headquarters further identified how bleak the electricity problems were in Basra. The fire at Hartha damaged the wiring. It would take several weeks to repair. General Strock made a strategic decision. He contacted the US Army Corps of Engineers' Prime Power unit to provide dedicated power for the refinery.

The 249th Engineer Battalion (Prime Power) is a versatile power-generation battalion that provides commercial-level power to military units and federal relief organizations. Each company in the battalion has four to six platoons capable of setting up, operating, and repairing power-generation and power-distribution systems worldwide. Each platoon has the capability to produce about three megawatts of power. That was how much power we needed to keep the refinery operational.

A platoon was in theater. It was immediately mobilized to a location adjacent to the refinery. They installed their generators at the power substation, feeding the refinery from the national grid. The US soldiers in the 249th Engineers were impressive. They worked hard and knew their business well. Power was flowing back into the refinery within a couple of days. The unit stayed on location for several months and the refinery received reliable power the entire time.

Thaer got his refinery operational, and adequate fuel supplies returned to Basra. News of the success of the 249th reached the oil ministry headquarters soon after I returned to Baghdad. When I walked into Ghadhban's office, he commented about this elite US Army unit that pulled into Basra and worked 24/7 until they delivered power to the refinery. He requested the same high-level support for his other two refineries if they lost power. It was another victory for the Corps of Engineers.

CHAPTER 10

The Bizarre Top Priority

Why is there so much interest in a pipeline that no longer exists?
—Oil Minister Thamir Al Ghadhban

Steve Coll, the Pulitzer Prize–winning author, was sitting across the table from me in 2008, interviewing me for his latest book on the international oil industry. He asked me to describe the most bizarre thing that happened to me during my first few months in Iraq. There were many challenging and difficult things that happened during the summer of 2003, but I had never been asked nor had I thought of something that was bizarre. Then one came to me, "It was the Haifa pipeline."

"What the hell is the Haifa pipeline?" Steve asked.

"Exactly," I said. "Those were my words when I got the call from a civilian at the Pentagon saying that the Haifa pipeline was their top priority."

It was early July 2003 at about 1600 Baghdad time, 0800 Pentagon time, when the only phone in the CPA oil office started ringing. It rang about eight times before I realized that it was our phone. The phone had just been installed earlier that day by the troops in the Signal Company supporting us with communications. I did not even know the phone number. I had made a mental note to ask our guys for the number and circulate it. The addition of a landline phone would really help our operation.

The phone was ringing. It was probably our signal troops checking the installation and functionality of the telephone. I got up from my desk and walked across the room to answer it.

"Hello, this is the oil office," I said.

I heard a familiar voice, but one that I had not heard since leaving the Pentagon in March. It was Michael Makovsky. It was a remarkably clear connection with just a slight overseas delay.

"Mike, this is a nice surprise. How did you ever get this number? The troops just installed this phone a few hours ago. I don't even know the number," I said.

Mike quickly got to the reason for his phone call and identified it as a top priority. He emphasized that his request was coming from the top of the Pentagon. He wanted to know the status of what he called the "Kirkuk-to-Hafu pipeline."

I had studied the oil infrastructure of Iraq for five months at the Pentagon and I continued while in Kuwait and Iraq. I thought I was familiar with every pipeline, oil field, refinery, and major piece of infrastructure, but I had never heard of the "Kirkuk-to-Hafu pipeline." I thought maybe I had misheard the name. "Is that spelled H-A-F-U?" I asked.

He let me know that my spelling was close enough.

Anyone who has lived in the Arab world knows that the English spelling of Arabic names is seldom consistent. For example, I have seen the city of Bayji spelled Baiji and Beiji in English. Also, I would learn years later that "Hafu" is close to the Hebrew pronunciation of the real name.

"When you say the top, are you talking about Doug Feith?" I asked Mike.

Mike let me know that the asker was higher than Feith, but he did not mention a name.

There were only two people above Feith at the Pentagon: Paul Wolfowitz and Donald Rumsfeld. Mike liked to drop the Wolfowitz name in the past, so I assumed it was coming from Wolfowitz. Now my curiosity was piqued. Why would the deputy secretary of defense, Paul Wolfowitz, be interested in a pipeline that I had never heard of? Did I overlook a key pipeline during my studies? There were rumors in the press about Iraq hiding WMD in pipelines to Syria.

"Does this have something to do with the rumors about WMD?" I asked.

Mike quickly responded that the phone line was not secure and so he could not discuss the reason for the request.

"Ok, I know where Kirkuk is located. Where is Hafu?" I asked.

He let me know that it was out to the west of Iraq, but he quickly changed the subject and said that he had to go to a meeting. He emphasized that he needed the status information quickly. Mike was in a hurry.

"I was just preparing for my Ghadhban visit the next morning. I have a list of items to review with him. This question is now on the list." I was eager to help.

Mike said that he would phone back the next day at the same time. I hung up the phone, impressed with the quality of the circuit. Having a landline with access to a good circuit back to the United States would be a marked improvement over the satellite phone and cell phone that I had been carrying.

Every day in Iraq felt urgent and important. Although there was rarely any personal danger to me, there were American soldiers deployed around the country who were depending on work that we were doing at CPA. They were in ever-present danger. Consequently, it was often difficult to sort and fully prioritize a daily list of operational and managerial issues. Everything felt like a number-one priority! Those priorities included the security of the oil infrastructure, the safety of our personnel, the recruitment of replacements, Iraqi payroll, the detention of oil ministry leadership, production bottlenecks, assessments of critical projects and funding, and on and on. All of these issues required attention daily if we were going to make the progress needed to eventually improve the lives of Iraqi people.

The next morning started like most of my days: the 0730 ministries meeting followed by the 0800 Bremer meeting followed by the 0830 oil sector meeting and then departure for the oil ministry around 0900 hrs. It was normally sometime around 0930 before I arrived at Thamir Ghadhban's office for our morning meeting. These meetings were at least three days a week.

Thamir Ghadhban and I would spend sixty to ninety minutes going through my list of current oil-related issues. Ghadhban would work through many of the issues during my visit and we would discuss the next steps for each specific issue before addressing the next item on my list. He would normally have to make multiple phone calls to help me resolve issues. Many of those concerns were requests coming in from our military commanders inquiring about more oil products for their area of operation. On this particular day, it took more than an hour before we got around to the last item.

"Thamir, I have one more item on my list. This one came in from the Pentagon yesterday afternoon. Can you get me the status of the Kirkuk-to-Hafu pipeline?" I asked.

When I looked up from my notes, I saw that Thamir got very grave and tensed up a little. "Gary, you do not want me to ask that question of this ministry."

I was a bit surprised by his reaction and immediately thought that maybe there was something to the rumor about hiding WMD in pipelines. I did not know what would make Thamir become so sober over my question.

"Ok, I'm game. Why should I not be officially asking the status of this pipeline?" I asked.

"Gary, there are a lot of people in this country and certainly a lot of people in the oil ministry who think that the primary reason that the United States came into Iraq was to take our oil and send it to Israel. If I officially ask the status of the old pipeline to Haifa, then I will be confirming their convictions."

Haifa, Israel, holy cow! I did not know such a pipeline existed and certainly did not realize that I was asking about a pipeline to Israel. This would obviously have some very sensitive implications. Makovsky's Hebrew pronunciation of Haifa obviously confused me.

"Thamir, please ignore that request and do nothing unless I get back to you. I did not realize that a pipeline to Israel had ever existed," I said.

"Gary, you have got to understand. You and I are engineers. We are motivated by design and economics. We make our decisions based on facts and logic. Yes, economically it might make sense for Iraq to have a pipeline that goes due west to the Mediterranean for our exports to the west. It makes economic sense to run the pipeline through any country along the way and pay tolls to that country. Unfortunately, Middle East pipeline decisions are based more on politics than economics, and I don't see that changing during our lifetimes."

Thamir was correct. I knew that. Pipeline decisions were always a result of a combination of economic, political, environmental, engineering, and financial factors. The issue of restarting a pipeline from Iraq to Israel might be an engineering and financial challenge. I did not know. But I did know for sure that it would indeed be a sensitive political topic. It was extremely complex and problematic.

Later that day, the new phone in the CPA oil office rang. Suspecting that it was Makovsky, I quickly answered. He immediately asked about the status of the pipeline.

"Mike, you didn't tell me that it was a pipeline to Haifa, Israel." No response was heard, so I continued. "Mike, this is a very sensitive issue. You need to help us a little more with this request. We need clarity on the background before proceeding. I need for you to put this request in writing explaining what you need and why you need it. I will have to review the request with Bremer's staff before proceeding. Once we get the local approval, we will go get you an answer."

There was a distinct clang and the line went dead. He hung up on me. I was stunned. I do not recall a professional employee ever hanging up on me. I certainly do not remember anyone ever intentionally hanging up on people during my private sector experience in the oil industry.

Keep in mind that our oil group would not know the information about Mike's background for another two years. It was not until 2005 that I became aware that Mike left the United States in 1989 to join the Israeli Foreign Service. For all I know, he could have been in the Israeli Foreign Service during the summer of 2003.[1]

The subject of Haifa never came up to me again. Our oil contact in the Defense Intelligence Agency (DIA) emailed me about Iraqi oil sales to Israel in late August 2003. "Was Iraq selling oil directly to Israel?" We knew the answer to that question. Since 1948, Iraq had refused to sell oil to Israel. Our DIA contact properly sourced that question as coming from Paul Wolfowitz.

Makovsky bypassed me and asked the question about the Haifa pipeline to Ghadhban by telephone a few days after our abbreviated phone conversation. Ghadhban informed him that Iraq had considered using that pipeline to deliver crude oil to a Jordanian refinery before 1991, but he had learned that the pipe no longer existed.

Mike is quoted in Steve Coll's book: "The Israelis were at one point interested in this at the beginning of the war," he recalled. "I was asked by an official to look at this." Mike investigated and wrote a brief report.[2]

Mike was not the only person interested in the Haifa pipeline during the summer of 2003. Two Israeli government ministers issued press releases announcing that the Haifa pipeline would soon be pumping Iraqi crude oil to Haifa.[3] Israeli Finance Minister Benjamin Netanyahu and National Infrastructures Minister Joseph Paritzky issued announcements in the press about receiving Iraqi crude oil before the end of 2003.

The following highlights the level of interest from elsewhere.

In December 2005, Thamir Ghadhban gave a talk at a think tank in Washington, DC, on the Iraqi oil sector. After his formal remarks, he took questions from the audience. The first question came from a young female reporter from *Reuters.*

"Mr. Ghadhban, why have you not yet turned on the flow of oil from Kirkuk to Haifa?" Several members of the audience that worked for international oil companies rolled their eyes or snickered.

Minister Ghadhban responded, "Before I answer your question, I have a question that I would like to ask you or anyone in the audience. I have been asked this question about the Haifa pipeline more than any other question about the Iraqi oil sector over the last three years. My question is, why is there so much interest in a pipeline that no longer exists?"

No one answered his question that day and, to my knowledge, no one has ever answered his question. He did answer the *Reuters* reporter's question, explaining that the Haifa pipeline was constructed by the British before World War II but stopped operating in 1948. Much of the pipeline in Anbar Province had been taken by the sheikhs and used to transport water.

Since that think tank presentation, I have learned many important facts that could have partially answered Thamir's question about why the Haifa pipeline was so important to some people. To use a military intelligence term, connect the following dots to better understand.

Dr. Ahmed Chalabi had promised the Jewish Institute for National Security Affairs (JINSA) in the United States in the 1990s that he would restore the Kirkuk-to-Haifa pipeline after the fall of Saddam. He made similar promises to other elements of the powerful Israeli lobby in the United States.[4]

Ahmed Chalabi was closely linked to Vice President Cheney's office. Chalabi had a secure (crypto) communications link from Baghdad to the VP's office in DC.

Some in Vice President Cheney's office, especially Lewis "Scooter" Libby, were closely linked to the government of Israel. Jack Straw, the former British foreign secretary, said, "It's a toss-up whether Libby is working for the Israelis or the Americans on any given day."[5]

Israel had to import 100 percent of its oil requirements in 2003. The Israeli Finance Minister Benjamin Netanyahu had raised the expectations of Israelis and Israeli supporters in the international community about receiving crude oil from Iraq in 2003.[6] The Israeli minister for national infrastructures, Joseph Paritzky, is quoted in the press as saying that Israel would be able to decrease the cost of energy imports by 25 percent once Iraqi crude oil was received in 2003. They were paying a premium to Russia for oil.[7]

A pipeline from Kirkuk to Haifa would be a huge benefit to Israel and a windfall to its investors. A large pipeline could transport one to two million barrels per day, $30 to $50 billion per year. Israel had lost its source of inexpensive Iranian oil in the 1990s. Iranian oil was transported through a secret pipeline between the two Israeli ports of Eilat and Ashkelon. This pipeline was completed in 1970 and was used to secretly transport Iranian crude oil that was marketed by Marc Rich. Rich was the successful oil trader who made billions of dollars on this secret pipeline through Israel. Most of his fortune in oil trading was made using this trans-Israel pipeline.[8]

Lewis "Scooter" Libby was the law partner of Richard Nixon's personal attorney, Leonard Garment. Under Garment's direction, off and on over a period of sixteen years, Libby was Marc Rich's attorney.[9]

Ghadhban explained to me years later that the interest in Iraq about this pipeline became so intense in the Iraqi press during 2003 that he had to write a written response at the end of that first summer. He had to explain in the Iraqi press that the Americans were not stealing Iraqi oil and shipping it to Israel. The pipeline was no longer operational and could not physically ship oil to Israel.

The perception of the United States sending Iraqi oil to Israel was a powder keg that exploded and contributed to the insurgency against the oil sector in the form of pipeline attacks.

Pipeline Attacks

The insurgency officially started in the oil sector during the third week of July 2003. My phone rang late one night. It was Ghadhban. "Gary, two crude oil pipelines have been attacked within a two-hour period. This is different from all the previous attacks. The others were primarily for economic reasons, to steal the oil. This attack is intended to shut down the Doura Refinery. It cuts off all supply of crude oil to Doura."

This was significant. Iraq had just recovered from severe shortages of petroleum products in May and June. Demand was now being met by refinery production, but there was no inventory build, no stocks. A shutdown of more than a day would translate into shortages, long lines at the gasoline stations, and instability.

The 0800 Bremer meeting the next morning was my opportunity to share the significance of this event. "Sir, we had an attack on the two crude pipelines feeding the Doura Refinery last night. The attacks were at two different locations, but occurred at about the same time. This will shut down the refinery for some period of time and shortages of products will be noticeable almost immediately. The sector needs help. We do not have the resources to secure these pipelines."

Ambassador Bremer looked grim. He spoke to Major General Carl Strock. Bremer directed that Strock get with Brigadier General Dan Hahn and me immediately following the meeting and report back later in the day with an update. General Sanchez was not in Baghdad at the time, so Brigadier General Dan Hahn was standing in for him.

Strock, Hahn, and I stayed after the Bremer meeting and I told them everything I knew. The facts indicated that it might have been an inside job. Only a very limited number of current or former employees of the oil ministry knew the routing of these pipelines. All maps had been controlled documents under the previous regime.

It was, in fact, an inside job. The director general of Doura Refinery visited the United States in late 2005 and shared details of this attack with me. He met with the only employees that knew the routing of these pipelines shortly after the incident. He asked them what they knew. One of them said that the Americans were sending Iraqi oil to Israel and he produced a newspaper article that quoted a London newspaper. "What was your response when you heard this?" I asked. The director general stated, "I chastised them. I said, 'You idiots, the Americans are not sending our oil to Israel. If they were, I would help you blow up our pipelines.'"

After the meeting with Generals Strock and Hahn, I immediately left for the oil ministry to meet with Ghadhban. He was still getting updates from his managers on the extent of the damage. The situation was dire.

"Gary, Dathar decreased production at the refinery. He only has a couple of days of crude oil supplies in his tanks. He hopes to avoid a total shutdown by decreasing the rate of production. The repair teams are still putting out the pipeline fire at the attack sites. It's going to take a while to repair the damage." Thamir was concerned.

"I noticed that the lines at the Galani service station were long. The word has gotten out to the street and the population is already starting to sense a shortage," I commented.

We sat and discussed our options. Brigadier General Bob Crear had been notified and he issued task orders to KBR for buying and delivering products as they did during the May crisis. Unfortunately, it would take a week or so to mobilize sizeable volumes and get those products into Baghdad. What were our other options?

Thamir had a good idea. "Gary, KBR can import from Turkey, Jordan, Saudi Arabia, and Kuwait. That is good. We have another neighbor to the west who might be able to help with this problem. SOMO has a history of working with Syria. How about SOMO take the lead on importing products from Syria?" SOMO was the State Oil Marketing Organization, responsible for exporting or importing all oil into Iraq. They had developed excellent contacts in Syria, as well as in Turkey and Jordan, while executing their smuggling operations before 2003 to bypass the sanctions regime.

There would be resistance to buying Syrian products. This resistance would come from inside the US government and most certainly from the senior civilian leadership at the Pentagon. "We definitely need the product and we need it quickly. Have SOMO check with their contacts to get an understanding of what products and how much they can supply. Do not make any commitments. This is sensitive and any decision to proceed will require approval from Ambassador Bremer," I said.

We scheduled a conference call that evening to update Phil Carroll, in London, and the two-person advisory board on the fuel crisis. Pam Quanrud had moved to another job at DOS, so Adnan Kuayat and Richard S. Karp at the NSC were now facilitating the calls. Mike Makovsky at OSD policy had also joined the last couple of conference calls since Pam's departure.

Phil welcomed everyone to the call and quickly asked Ghadhban to provide an update on the status of shortages and the current plan to fix the problems. Thamir went into the details on locations of product volumes and the distribution to specific areas with acute shortages. He reviewed import supplies committed by Task Force RIO and KBR and an estimate on the return of production of the Doura Refinery. As an afterthought, he said, "Gary, SOMO contacted the Syrians today about product sourcing and—"

At that point, Makovsky interrupted and yelled over the phone. He shouted that Ghadhban could not contact the Syrians, that he could not talk to the Syrians, and that he would not talk to the Syrians.

There was a long pause when Mike stopped his tirade. Then, Phil spoke up. "Well, that's all for this evening. We'll talk again next week." I hung up my phone and sat there thinking; the arrogance of Makovsky was unbelievable.

Not more than fifteen seconds passed before my phone rang. It was Phil. "Gary, that was the last conference call for this group. No more. The official reason is that the calls are no longer needed. The real reason is that I will not tolerate a low-level staffer, sitting in a safe and secure refuge, insulting a respected Iraqi oil executive like Thamir. Especially as we sit here on the brink of fuel shortages and a widespread panic."

Phil returned to Baghdad at the end of July for a short time before heading to a CENTCOM meeting in Tampa in early August. The shortage of products was still present in Baghdad. Long gas lines were all over the city. Any insurgency group could see the effectiveness of attacks on oil pipelines. Most of the vehicles inside Baghdad were in lines at one of the gas stations. Daily temperatures in Baghdad were very hot. People were growing more and more frustrated. Attacks on oil pipelines by the insurgents became a common occurrence after the instability created by this first attack.

The thirty to fifty trucks a day that Syria could have provided would have been a big help, but Phil refused to consider it. He said that he refused to embarrass the administration. I did not understand his comment, but he let me know that he did not want to discuss it. I suspect it had something to do with the administration's efforts to get Congress to vote for economic sanctions against Syria, and the following story supports my position.

In August, Major General Petraeus asked for my support in allowing the 101st to exchange crude oil in northern Iraq with Syria for petroleum products. One of his soldiers had been killed while trying to control the crowds gathered to buy scarce products. The general needed products.

My orders from Phil were very clear. I could not support the Petraeus request, but my stomach was in knots about the soldier's death. The only guidance that I could give was to remind Petraeus of our West Point plebe-year lessons on the responsibilities of the commander to protect his troops. That responsibility gave him more authority than I could get him.

Reports came into the office a couple of weeks later that the Iraqi governor in the north had signed a deal with the Syrians to provide crude oil to Syria in exchange for the delivery of thirty to forty tanker trucks a day of products. The smile on my face was pretty obvious. The person giving the report insisted that CPA oil take action to stop the sale. "I have no authority over governors," I said and walked away. I had no energy to stop a governor's action that I knew was saving the lives of US soldiers.

The exchange deal with Syria was quite clever. There was a small (eight inches in diameter) export pipeline near Sufaya, northwest of Mosul, for exporting crude oil to Syria. The 101st had already instructed the local oil ministry management to ship a small amount of crude oil across the border to Syria in exchange for electric power. This cross-border exchange for electricity had been going on prior to the war. Ambassador Bremer gave Steve Browning (the electricity advisor) and me the approval to help the 101st negotiate exchanges with Turkey and Syria to get the electricity.

There was unused capacity on the pipeline and there was plenty of capped production capacity in the oil field at Ayn Zalah. The production management of the local oil ministry in Ayn Zalah had the capability of shipping much more crude oil to Syria than they were exchanging for electricity. The governor had signed a deal to send this additional crude oil in exchange for gasoline and diesel. It was a win for everyone. The Syrians got crude oil, the Iraqis in the Mosul area received products, and the 101st experienced a decrease in attacks on soldiers guarding the gasoline lines.

Not quite. There was an apparent loser, our civilian leaders back at the Pentagon. The phone in the CPA oil office rang at 1600 hours. It was Mike Makovsky at the Pentagon.

Mike asked me if I gave permission for the governor to sign an oil deal with the Syrians.

"No, Mike," I responded.

Mike challenged me by asking who gave the governor permission.

"Mike, I don't know." I suspected that Petraeus had directed the governor, but I did not know and had no desire to investigate.

Mike had heard a rumor that Petraeus may have given the governor permission and he asked me to confirm it.

"Mike, I told you that I do not know who authorized the governor."

Then Mike showed his irritation with my response and raised his voice. He said that if he confirmed that Petraeus directed the governor, Petraeus would be one sorry general. Mike's arrogance was coming out. I tried to calm him down.

"Mike, please come out here to understand—" I heard a clang and the phone went dead. That was Makovsky's last phone call to me. He avoided talking to me in the future when he needed information.

The *Early Bird* was the popular roundup of global news reports on military, defense, and security matters. It was available to anyone with access to the DOD computer system. It supplied a short summary of news articles from many sources with the capability of drilling down and reading the entire article.

An article showed up in the *Early Bird* quoting a congressman. It was about a week after Makovsky's threat. The congressman was demanding that the Pentagon fire Major General Petraeus for trading oil with the Syrians. My immediate reaction was that Mike's threats must have been coming from someone higher up in his chain of command to yield that amount of power.

Someone with good access to Capitol Hill was involved. Who? I did not know then but would come to speculate that it was his AIPAC network.

The attack showed up in the *Early Bird* only one time. It had no legs, no follow-up. It may have been outdone by an article that ran in the *New York Times* on September 4, 2003, by Michael Gordon praising General Petraeus and the 101st for their great job establishing cross-border trading with Syria of several goods, not just oil.[10]

Some congressmen started challenging the wisdom of their own work in Congress, but the Syrian sanctions passed with HR 1828. "Such sanctions will not help alleviate the incessant attacks that our soldiers are facing daily in Iraq," Honorable Nick Rahall II of West Virginia stated on the floor. He entered an article from the *Wall Street Journal* into the Congressional Record that identified the disconnect between the reality of what was happening on the ground in Iraq and what Congress was pushing through as legislation: "IRAQ ADDS COMPLEXITY FOR U.S., SYRIA AS WASHINGTON SANCTIONS DAMASCUS, AMERICAN TROOPS SEEK SYRIAN TRADE PARTNERS."[11]

The *New York Times* and the US Army had Petraeus's back on this incident. As a civilian with no political sponsors, no one had my back. This was the beginning of an effort by Makovsky and the neoconservatives to discredit me. I received a "heads-up" email from Richard Greco on August 25 advising me that the Syria issue had been discussed at a recent Deputies meeting in Washington. Someone at the meeting erroneously mentioned that I gave the "OK to proceed" concerning the oil trade with Syria. Richard was a political appointee who had worked in Baghdad with me earlier in 2003. He and I had gotten along quite well, but he would soon be leaving his Pentagon job. I emailed Richard and copied Reuben Jeffery (the director of CPA West) that my work was part of an initiative authorized by Ambassador Bremer to negotiate with Turkey and Syria for electricity. Unfortunately, no one in Washington was willing to defend me. But then, no one should have to be defended for trying to save the lives of US soldiers.

Any trade with Syria was a "red line" with the neocons. Dr. Harold Rhode had shared with the ORHA team earlier in 2003 that the grand plan of the Pentagon civilians was war with Syria and Iran, after Baghdad fell.[12]

It would be held against me that I participated in this initiative to deliver Syrian fuel to Mosul to save the lives of US soldiers. Many friends would advise me that Makovsky blocked or attempted to block any of my future oil requests or initiatives that went through the CPA West or OSD policy offices at the Pentagon. I was advised that he blocked me from jobs in 2005 and again in 2006.

The inquiries that I received from Makovsky during the summer of 2003 concerning the Haifa pipeline were bizarre. The intolerant predisposition of the neocons about trading with Syria for electricity and products was counterproductive at best. Both were distractions to our mission and probably cost American lives. It is naïve to believe that Makovsky was a standalone eccentric, as one Republican political appointee at the Pentagon wanted me to believe. Another political appointee referred to him as being part of "Wolfie's gang," referring to Paul Wolfowitz. He definitely had powerful sponsors. It became evident in 2005 that he probably had excellent contacts in the Israeli government. He was definitely working on a different agenda than the rest of us were and must have done it well; he was appointed CEO of the Jewish Institute for National Security Affairs (JINSA) in 2013.

We lost another forty-eight military personnel in July and thirty-five in August, bringing the total since March to almost three hundred. We had no idea in the summer of 2003 that the killed in action (KIA) number would get to 4,489 by the end of 2011.[13]

CHAPTER 11

Task Force Shield

Securing the Oil Sector

Find a good colonel.
—CPA chief of staff

The need for security in the oil sector became apparent soon after we entered Iraq. The looting at oil facilities throughout the country continued during June 2003. The South Oil Company (SOC) headquarters in downtown Basra was looted twice during the first couple of months. The South Gas Company, located west of Basra, was severely looted after KBR did their initial assessment in May. Any oil facility was looted that did not have the physical presence of US or Iraqi security personnel twenty-four hours a day. The ministry headquarters in Baghdad was the only oil facility that housed a US military unit.

Oil sector security under Saddam fell under the Ministry of Interior along with many security functions such as city police and all other security for the various ministries. ORHA and CPA Ministry of Interior advisors let us know that it would be at least one year before they could get around to building a security force for the Ministry of Oil.

We went to General Sanchez in May 2003 to request security from the coalition forces because of the looting. Also, dedicated security was needed for the convoys of tanker trucks importing products into the country. Sanchez only agreed to provide security for the convoys, until contract security was obtained. There were not enough military resources to provide fixed-site security. We were instructed to get our own security, to contract it out. We were on our own for finding oil infrastructure security.

The CPA oil team immediately looked to the Corps of Engineers and KBR to expedite a security task order under their contract. The contract with KBR offered us the only realistic mechanism to expeditiously obtain security. KBR already had limited security resources in the country supporting their oil reconstruction work. Using KBR became troublesome for retired Admiral David Oliver, the CPA budget director. He was sympathetic to political complaints about Vice President Cheney's old company, KBR. He refused to allow us to use the KBR single-source contract for a security tasking. Instead, we were given priority to expedite a security contract through the CPA procurement organization under an invitation to bid.

Even when a government procurement action was given top priority in a war zone, it still took three months. The looting continued.

The CPA oil team chose Major Rebecca Freeze and Colonel Gary Major to lead the effort. They developed the statement of work and coordinated with the CPA procurement office. Thamir Ghadhban assigned two senior oil executives from the Ministry of Oil to work with our team, Natiq Bayati and Abdulilah Al Amir.

Natiq was educated in the United Kingdom and would become well known in 2009 in the job of director general of the Petroleum Contracts and Licensing Division (PCLD). PCLD led the ministry's bid rounds for the return of international oil companies to Iraq. Abdulilah was also educated in the United Kingdom. He would be a key advisor to both Thamir Ghadhban and deputy prime minister for energy Shahristani from 2008 to 2011.

The team mapped out the locations within the oil sector that required physical security. These locations were included as part of the invitation to bid. They required a security force of a little less than ten thousand guards.

Recognizing the importance of getting a security force in place as quickly as possible, Phil Carroll assigned this effort our top priority and monitored the progress of the procurement action as best he could. It was a public procurement action, so we had to be careful to follow the guidance of the Federal Acquisition Regulations. Even with the priority and our daily focus on the need to get urgent security for the oil sector, the process still took us three months.

A contract was awarded in August 2003 to an unknown small group of former UK army officers who formed a security group called Erinys. Erinys had the financial backing of Dr. Ahmed Chalabi. They were selected primarily because of a well-written plan and a low bid.

A request was made of CENTCOM to assign a brigadier general to lead the contracted security effort. CENTCOM did not give the request any priority. When it became obvious that military channels were moving too slow, we approached Ambassador Patrick Kennedy, Bremer's chief of staff. Pat gave me great guidance that should always be considered for anyone working in a war zone. He said that general officers were high maintenance in a war zone. He told me to find the best colonel in Baghdad. I was surprised at Pat's candidness, but he was absolutely correct. General officers required an aide-de-camp, staff, and special handling in a war zone. We needed an action-oriented colonel.

As General Marshall once said, "I want an officer for a dangerous mission, I want a West Point football player!" We were able to do better. We found a West Point lacrosse player who was number one in the class of 1979 for physical education. Tom O'Donnell had just arrived in early July and had been assigned as the chief of staff under Major General Paul Eaton, who was leading the effort to build the new Iraqi army. Paul was building a large staff of senior US military officers and Tom was his chief.

O'Donnell impressed Kennedy during an impromptu interview. Tom had been a successful brigade commander, commanding the US Army Training Center for Field Artillery at Fort Sill, Oklahoma. He knew nothing about oil, but he impressed us with how he felt that oil facilities should be secured, similar to the way an army artillery commander secures his location. A good commander performs a reconnaissance of his routes of movement before relocating his battery of howitzers. Oil pipelines should be handled the same way—"reconn'd" and secured.[1]

Pat Kennedy and I negotiated a deal on August 7 with Paul Eaton to give up O'Donnell in return for the next senior colonel arriving in CPA. Paul was not happy about giving up Tom but recognized the importance of our mission, and he knew that Tom could handle the challenge. To help ease the loss, I gave Paul a couple of cold sixteen-ounce cans of Heinekens, a very scarce commodity in Baghdad during the summer of 2003.

Tom was thrown into the fire on his second day on the job. A meeting was held at the ministry on August 8 with Erinys, O'Donnell, Ghadhban, and myself. It was the kickoff meeting for Erinys and their introduction to Thamir Ghadhban. Erinys brought along a well-known retired UK general officer to assist them and lend credibility to their presentation.

The meeting did not go well. Erinys was working off of a statement of work developed three months prior. Since then, the most serious threat had moved from looting oil facilities to attacks on pipelines. Erinys was responding to a contract issued by the US government that was now overcome by events (OBE). They were responding to the contractual requirements and not to the fluidity of the battlefield. Many things change in a war zone over a three-month period.

Ghadhban was frustrated that this expensive contract of about $40 million would not meet his immediate threat. This money was coming from the sale of Iraqi oil and not the US government, so Thamir rightfully accepted his role of properly stewarding the disbursement. Tom and I convinced Thamir to be patient for a week. We would return with a plan to address the more immediate threat to oil operations.

Our plan was to amend the Erinys contract, adjust the scope of work, and increase the number of Iraqi guards required to secure the key facilities and pipelines. Follow-up meetings between Colonel O'Donnell and Thamir Ghadhban brought the contractual obligations of Erinys in line with the operational needs of the oil ministry to secure the country's oil infrastructure and do it quickly.

Ghadhban inserted himself into managing this effort. He and Tom met several times to complete the task. Security was too important and time was of the essence. Ghadhban identified every oil ministry location in the country. There were more than seven hundred locations. He and Tom narrowed down and prioritized the sites to a more manageable 256 locations that required immediate security. He also highlighted to O'Donnell the key pipelines that required immediate security and those pipelines were given the highest priority. Ghadhban suggested that O'Donnell engage the tribes to secure the pipelines in their tribal area throughout the country. This suggestion proved to be very effective but difficult to execute.

Determining where one tribal area ended and another tribal area started was seldom easy. O'Donnell had to get help from the CPA tribal experts to ensure he had exact grid coordinates of the tribal areas.

O'Donnell came up with the idea of calling this new oil security unit a military name, Task Force Shield. Branding his unit worked very well and the brand soon became well known within Iraq. We discussed several names, but Shield was the most appropriate.

Using the title of "Task Force" created some initial confusion at the senior levels between CPA and CJTF-7. All military organizations were supposed to report to General Sanchez and CJTF-7, but this one initially fell under CPA.

The name proved invaluable when working with the various coalition commanders throughout Iraq. O'Donnell got the CJTF-7 operations section to issue General Order #21, which announced the existence of Task Force Shield to all the coalition military units. The order explained the mission of Task Force Shield and identified their points of contact. It identified Colonel Thomas O'Donnell as commander. O'Donnell's reputation was well known among field artillery officers such as Major General Ray Odierno, commander of the 4th Infantry Division based in Tikrit.

One of O'Donnell's previous jobs was as the assignments officer for the field artillery branch (colonels to lieutenants) at the Army Personnel Command (PERSCOM). He held this position in the late 1990s and managed the assignments for many of the field artillery officers now in key command positions throughout Iraq. He was well known within the artillery branch and the other combat arms branches by his nickname, "OD."

O'Donnell put together a plan to visit all of the oil sector sites throughout Iraq. During September 2003, he traveled from Dahok in Kurdistan near the

Turkish border to Basra in the south near the Kuwait border. This was the first of three trips throughout Iraq, which each lasted up to two weeks. He inspected pipeline routes and static oil infrastructure sites.

Tom's personal security team consisted of six burley, heavily armed South African guards traveling in three black Suburbans, initially not armored. The vehicles were equipped with Thuraya phones and military and commercial communication equipment. They were eventually connected to the Eriny's countrywide VSAT system. In addition, they carried large supplies of ammunition, MREs, bottled water, diesel gas cans, and ample medical supplies.

Local translators were a major part of the coalition's success and this was the case for Task Force Shield. Tom selected Dahir Al Nadar. He was an Iraqi raised in Nasiriya with a bachelor's degree in computer science. He was thirty-two years old and spoke fluent English, French, Arabic, and Kurdish (a rare skill). Dahir traveled everywhere with Tom and was a valuable resource at meetings with sheikhs and tribal elders, community engagements, tribal disputes, and ad hoc encounters with Iraqis and Kurds. But equally important were the private conversions they had while traveling in their vehicle. Tom valued the companionship and cultural education that Dahir shared with him.

Dahir helped Tom to quickly adapt to the Arabic culture and customs. He proved invaluable. Getting Task Force Shield operating effectively could not have happened without Dahir.

One morning in early February 2004, there was a suicide bomber attack just outside of Assassins' Gate, one of the main entrances to the Green Zone and CPA headquarters. Everyone in the Green Zone heard and felt the explosion. There was a long line of Iraqis who were CPA employees awaiting full-body searches and identification checks before entering the grounds.

Just after the explosion, Tom received a call from Dahir saying he would be late coming to the office because of a problem at the gate. When asked if he was OK, Dahir calmly said, "I am blown-up sir, but fine." He was thirty meters away from the bomber's car. During the explosion, Dahir was blown ten feet into a wall, ruptured his right eardrum, fractured his shoulder, and had cuts on his face and arms. Four hours later, he showed up at the office ready to work.

O'Donnell's first countrywide trip was an assessment, taking inventory. This is what any new commander does. What he found was quite disturbing. There was essentially no guard force in place. Tom found one guard with his cousin in Kirkuk. The guard said he guarded the strategically important water plant for the water-injection system of the Kirkuk oil fields. The two guards lived in a little guardhouse full time and kept an AK-47 locked up in a safe for fear that someone might steal it. They wore no uniform other than their traditional long, white male dishdasha.

Tom's second visit to oil sector sites followed quickly, within a month. It was a working trip to establish nine headquarter locations where Erinys would staff each location with ex-pat personnel. Tom made contact with the closest local coalition commander to ensure that the Task Force Shield headquarters could be logistically and tactically supported. This support consisted of food, fuel for vehicles, and generators, as well as quick reaction assistance to help in any major attack of the oil facilities or pipeline. The coalition commanders also helped with acquiring buildings for supporting the Shield mission.

During these trips, Tom briefed coalition division commanders. A few of the most recognized names were Brigadier General Martin Dempsey, Major General David Petraeus, and Major General Ray Odierno. He met with many of the sheikhs and signed contracts for their support in securing the pipeline in their area or in providing guards at the oil facilities.

Bringing in a new Iraqi recruit to the Shield organization required a minimum of basic training. They were put through a five-day training program. The recruits were issued uniforms and a badge. This proved to be extremely important. It provided a respectable job to many young male Iraqis. The guards were paid in (US) cash. Hiring young men from the tribes won over the support of many of the powerful sheikhs. The job of a security guard was a good job.

The young Iraqi men knew how to handle an AK-47. They had experience with military organizations and were sharp enough to pick up other skills. The training focused primarily on conducting vehicle searches, using communications equipment, and learning the General Orders taught to all US military recruits. General Order number one was highly emphasized to every Iraqi, just as it is to every American soldier.

> Army General Order #1—I will guard everything within the limits of my post and quit my post only when properly relieved.

O'Donnell's last trip through Iraq was in the spring of 2004, after most of the guard force was in place. This visit included an inspection of the Iraqi guards in their blue uniforms, weapons, fortifications, and coordination with the local Ministry of Oil management. Many of the senior Iraqi recruits were former senior military members in the Iraqi military.

Tom witnessed what a job meant to the fifteen thousand Iraqi members of Task Force Shield. These Iraqis took pride in getting a job, a serious and respectable job. This was extremely important. Task Force Shield took fifteen thousand potential insurgents off the battlefield during a time when

Colonel Tom O'Donnell visits his Task Force Shield guards in their new uniforms at the Bayji Refinery in October 2003. Photo courtesy of O'Donnell.

unemployed former Iraqi soldiers were joining the ranks of any organization willing to pay them. Training on the communications equipment and how to contact the local coalition commander was important, but the uniform and a badge provided instant respect for the Iraqi guard in his community. That Task Force Shield security guard was guarding Iraq's national treasure, oil.

O'Donnell held formal formations at many locations and recognized the best Iraqi members in front of their peers. He presented the highly coveted Task Force Shield coin, a commander's coin that Tom designed and procured shortly after his new assignment.

Using mobile patrols became the best way to secure the pipelines both in tribal areas and in areas not controlled by specific tribes. Task Force Shield purchased about five hundred vehicles to support this effort. These were normally thin-skinned pickup trucks or small vehicles.

Immediately following the contract award, Erinys established an operation center in the Mansur District of Baghdad. They set up proper communications equipment, vehicles, and other operational logistics. Their communications equipment enabled them to have radio communications throughout the country. Erinys brought in ex-pat recruits from the United Kingdom, South Africa, and North America with military experience. They organized the country into regions and set up regional managers with headquarter offices in those areas responsible for training, uniforms, logistics, and paying employees.

Task Force Shield proved to be one of the successes within the oil sector during 2003 and 2004. The Shield organization delivered on their commitments.

O'Donnell scattered his US military team throughout Iraq, colocating them at the Erinys' regional headquarters. They became Tom's eyes and ears in the field. Some of those team members included Lieutenant Colonel Tony Cusimano, Lieutenant Colonel Matt Croke, Captain Scott Davis, Captain Morgan Ashton, Sergeant Jared Jalbert, Sergeant Thomas Shedd, and many others.

The reputation of Task Force Shield grew within the Ministry of Oil leadership. Colonel O'Donnell had regular meetings with Thamir Ghadhban to ensure that he stayed aligned with the ministry's objectives. Anecdotal stories of specific attacks on oil infrastructure locations that were repulsed by Task Force Shield trickled in from the field. The story that best summarizes the success of Shield was its defense of the Al Basra oil terminal (ABOT).

The coalition navy pulled off its forces in early 2004 from the export terminal in the south. Several guards under Task Force Shield replaced the coalition navy on the terminal. These guards were armed with light weapons and stationed on the terminal, fifty kilometers offshore. Two speedboats loaded with explosives and driven by suicide bombers approached ABOT on the evening of April 24, 2004, at high speed.[2] The Shield guards on the terminal opened fire, forcing the speedboats to prematurely detonate. The US Coast Guard engaged a third boat that was speeding close to the smaller terminal, KAAOT. Two US sailors were killed when the boat detonated its explosives.

A successful attack on the Al Basra terminal in early 2004 would have been catastrophic. ABOT represented more than 90 percent of the revenues coming into Iraq at that time. This near catastrophe forced CENTCOM to rethink its mission and to return coalition navy resources back to the terminals, where they remained until the end of 2011.

Our main concern with an Iraqi oil security force in uniform and carrying AK-47s was a "blue-on-blue" incident. Our US soldiers had just been shooting any suspicious Iraqis carrying a weapon a few months prior. Now, we were arming friendly Iraqis to secure the oil infrastructure in areas controlled by coalition forces. Tom showed pictures of his Task Force Shield guards to all of the commanders he visited: general officers, colonels, and captains. He wanted to get the word out to the coalition soldiers that these were friendly Iraqis operating in their area of operation.[3]

The contractor, Erinys, had been charged with increasing their ranks to a number close to fifteen thousand Iraqis as soon as possible. The risk of a blue-on-blue incident increased as the number of guards increased.

Our nightmare became a reality in late October 2003. It occurred on the perimeter of the North Oil Company in Kirkuk. It was a very cold night and the Iraqi members of Task Force Shield augmented their blue uniforms with additional civilian clothing to stay warm. Their heads were wrapped with scarves that covered their blue hats. They wore additional jackets over their Task Force Shield blue shirts as they were out on patrol on the perimeter. There were seven security members, all carrying AK-47s.

A US patrol in Humvees approached the location and shined lights on the Shield security team. It was dark, no moon and no artificial lighting. The Iraqis panicked, some dropped their weapons and ran. Some ran in the direction of the US forces and some ran in other directions. The Americans opened fire, hitting all seven.

The on-site American leader soon realized that his team had fired on friendly Iraqis. The wounded Iraqis were quickly moved to the local American medical facility for treatment. Three of the Iraqis died and four survived this tragic incident.

The local battalion commander called for a 15–6 investigation of the incident. Investigations under Army Regulation 15–6 are administrative fact-finding investigations by a military officer who is independent of the accused unit. All the facts were considered: pitch-black darkness, Iraqis in no recognizable clothing, AK-47s at the location, and Iraqis running when the American unit saw them. This was truly the result of the "fog of war." The investigator concluded that it was an unfortunate accident.[4]

The leadership of Task Force Shield used this incident as a catalyst to implement new procedures for the Iraqi guards. It was emphasized throughout the country that anytime Task Force Shield Iraqi guards were approached by coalition military they should immediately freeze, place their weapons on the ground, and put their hands in the air. The incident in Kirkuk was fortunately the only fatal "blue-on-blue" incident.[5]

Traveling around Iraq with three black SUVs in 2003 and 2004 made Colonel O'Donnell and his small personal security team a target. Tom was involved in more direct fire attacks than any other colonel in Iraq at the time.

The first incident occurred just west of Kirkuk, when his convoy received direct fire from vehicles going in the opposite direction. Fraser Brown, a senior Erinys employee, was hit and required immediate medical attention. The rest of their trip to Kirkuk was at a very high rate of speed. Arriving at their destination, the team moved into the Kirkuk Air Force Base to receive medical treatment from the US forces. Unfortunately, an attack had just occurred at the base. So, it was on lockdown upon their arrival. No lights allowed, and it was after dark.

Colonel O'Donnell had to signal his way through the perimeter and to the medical facility, using orange and yellow "glow sticks" to indicate his route of travel. Tom opened his SUV back door and leaned outside of the vehicle. He held the glow stick in one hand and securely grasped the vehicle with his other hand. Fraser was lying across the back seat. He was immobile from his injury. The person in the front seat stayed on the radio with the base operations center to ensure that everyone understood their route as they moved to the medical facility. Fraser was treated and later released back to limited duty.[6]

The second incident occurred outside of Fallujah in Anbar Province. It was January 2004 and O'Donnell's personal security team had just exchanged their soft skin SUVs for armored SUVs. Their timing was opportune. The attack in Fallujah was an ambush that sprayed bullets at all their vehicles. One of the bullets hit the door where Tom was seated and was only stopped by the armor plate recently installed.

The third incident took place in an area located southwest of Baghdad. We called it the triangle of death. The vehicles were traveling through a dangerous area and were suddenly ambushed. As in any ambush, the situation was extremely fluid. Vehicles sped up to get through the kill zone when a flash of light was seen in the front and then an explosion blew up a building behind Tom's SUV.

Upon arriving in a safe location, the driver of Tom's vehicle inspected his SUV and found burn marks and a scrape on the roof just behind the windshield. The conclusion was that an RPG rocket had apparently glanced off the roof of the vehicle before destroying the building behind them.[7]

Probably the most violent attack on Tom's vehicles came as a result of a friendly fire incident. His SUVs were moving into the CPA compound located in Hilla, about an hour southwest of Baghdad. A new coalition force had just been assigned to secure the CPA compound. It was a unit from the Czech army.

It was late in the day, just before darkness. Tom's SUVs approached the compound entrance for the night. They had been to this compound many times in the past, but this time something had changed. The road approaching the compound had previously been a two-way road but had recently been changed to one-way, exit only. Tom's driver did not recognize the change until it was too late. The Czech guards opened up with .50-caliber machinegun fire. The .50-caliber rounds hit the front of the vehicle and the concrete barriers on both sides of the SUV. The driver stopped and remained still while Tom opened his back door with his hands in the air, waving his colonel's hat with the silver rank insignia. He shouted, "US military, US military," to the Czech guards.

The US Military Police assigned to the Hilla CPA site heard the shots and quickly arrived on location. They approached the vehicles. Tom's hands were still in the air. He was eager to prove Task Force Shield's identity, shouting his identity to everyone in the area. The situation calmed as identifications were verified. Tom took the opportunity to make his on-the-spot correction of the Czech guards by venting his anger over the near-fatal incident.

The new Czech guards were chastised by the chain of command for being so trigger-happy. They resumed their duties and chalked it up as another incident of the fog of war. Black SUVs with antennae would stand out to anyone that had been in Iraq for any length of time. Such vehicles driven in Iraq were almost always Americans. Being new, the Czechs felt like a vehicle trying to enter the exit route must be unfriendly. It was a close call.[8]

As the number of Shield guards and support systems increased, the number of attacks on the pipelines decreased. There were seventeen attacks on the oil pipelines in September 2003. The number decreased to three in May 2004. The months in between showed a steady decrease for each month. As Shield increased in size, attacks decreased. Likewise, as the attacks on the pipelines decreased, the ability and speed of the Ministry of Oil to repair the breaks dramatically improved. With only two or three attacks per month, the repair time decreased from weeks to days.

Unfortunately, O'Donnell's departure in the summer of 2004 saw an increase in attacks on pipelines. Attacks on pipelines would become a successful strategy of the insurgency for the following three years. Very little oil flowed through oil pipelines north of Basra and especially north of Baghdad during the violent years of 2006 and 2007.

The oil security force was eventually placed back under the Ministry of Interior in 2007, as it had been during previous regimes. An Iraqi military officer, General Hamid Abdullah Ibrahim, was appointed in 2007 and was still commanding the oil security force in 2015. His offices were located inside the oil ministry's walled complex on Portside Street, the old US military "Route Irish." The force grew to more than thirty thousand Iraqis. Their mission remained the same as Task Force Shield's—to protect Iraq's national treasure, oil.

CHAPTER 12

Changing of the Guard

Oil Ministry and CPA Oil Changes

Ambassador Bremer, I cannot accept the job of minister of oil.
—Thamir Al Ghadhban

As the summer of 2003 came to an end, the recently appointed Iraqi Governing Council met to appoint their new cabinet of ministers, a group of twenty-five leaders of Iraq's ministries. The appointments became effective September 1, 2003. The selection of an oil minister was not routine.

Ambassador Bremer approached Thamir Ghadhban in early August about being the minister of oil. Bremer had observed Ghadhban during the previous three months and was impressed by what he saw. He knew that the oil sector was the economy; 95 percent of Iraq's revenues depended on the oil sector. Bremer wanted to continue with Ghadhban in the leadership position. Ghadhban asked the ambassador if he would give him twenty-four hours to think it over and get back to him. Ambassador Bremer was surprised at the request but agreed.

The next morning I arrived at the ministry headquarters and walked into the office of Dheyaa Al Bakka, Thamir's office manager. Dheyaa was distressed. He looked at me and said that I needed to talk Thamir into taking the oil minister job. He did not want to take it, but Dheyaa asked that I convince him to accept it.

Well, I guess I knew the answer to the question before I even asked it. I walked into Thamir's office, sat down, and asked anyway, "What's your decision?"

Thamir looked at me and said, "Gary, I cannot accept the position of oil minister. Please thank Ambassador Bremer for his show of confidence, but tell him that I cannot accept."

"Oh no you don't. I am not delivering that message. That's a message you will have to deliver yourself." I was very disappointed. I looked at him and asked, "Why?"

His response, "Gary, I am exhausted. You've seen how busy we have been this summer. It has been nonstop, especially the last two months. I have not seen my oldest son in months. I need a break."

"We can get you a break. Take a week or two off. We can find someone to fill in temporarily during your absence," I said.

It had been an exhausting summer. I had taken a week off at the end of June to go back to the States. Thamir had been going 24/7 since late April and the pressure on him had been significant. On top of the workload, his name recently appeared in a newspaper issued by the Sadrists calling on their people to assassinate several Iraqi leaders of the previous regime. Thamir's name was among those targeted for assassination.

"Gary, it is not just the workload. I don't think I could work with many of these people that Ambassador Bremer has appointed to the Governing Council. There are many ex-pat Iraqis who are just opportunists that you Americans have given more credibility to than they deserve," he said.

"Thamir, for what it's worth, I think you are making a huge mistake. Just talking to Dheyaa outside, he thinks you are making a huge mistake. Please, rethink this thing through," I said.

His mind was made up. We met later that afternoon for an appointment with Ambassador Bremer. I had warned the ambassador before our arrival, so it was no surprise to him when Thamir said, "Ambassador Bremer, I cannot accept the job of minister of oil." The ambassador noted Thamir's comment and only requested that he not tell anyone that he had turned down the job.

Ambassador Bremer met with the Governing Council the next day to select the key members of the new cabinet. He did not tell anyone that Thamir had turned him down. Instead, he went into the session with the assumption that all ministry positions were open, except for Oil. He pretended that Thamir Ghadhban would be appointed as the oil minister. The members of the Governing Council were furious. Oil was the prize. That was the most important ministry in many of their eyes.

Mohammed Bahr Al Ulloum was one that was more furious than the others. He wanted his son to be the next oil minister. Ambassador Bremer negotiated the oil minister position away in return for a couple of other key ministry positions where he wanted competent people. Bremer essentially negotiated away something that he never had. It was a wise move on his part.

Ibrahim Bahr Al Ulloum was appointed the new minister of oil in early September. He entered the ministry complex a couple of weeks later. Thamir and I waited on the front steps of the ministry for Ibrahim. He arrived late in the day, about 4 PM. The three of us walked to the minister's new office on the fourth floor of the ministry complex.

Once inside the minister's office, the three of us sat down for a brief discussion. I looked at Thamir and asked him to give me five minutes with Bahr Al Ulloum and then I would leave for the day. Thamir left the room and closed the door.

"Your Excellency, when I saw you in Washington, DC, last January, did you ever think that we would be sitting here like this nine months later?" I asked.

Bahr Al Ulloum said that he knew all along that he would get the oil minister job. He mentioned that it had been predetermined. He did not say anything more to me, but I was curious. Why was he so confident that he would be the minister of oil? What did he know that I did not know? I would eventually learn that Dr. Ahmed Chalabi backed Bahr Al Ulloum. He was Chalabi's man. This was part of an agenda that I was not privy to.

"Phil Carroll sends his regrets for not being here. He had to attend a meeting with Ambassador Bremer, but he wants me to pass along this message to you. He and I have been impressed with the talents of Thamir Ghadhban. He is a world-class oil executive. He is an asset to this ministry, highly respected, and admired by everyone that has worked with him over the last few months. We highly recommend that you leave the day-to-day operations of the ministry to Thamir, your chief executive officer. We recommend that your role as oil minister have more of an upward focus on the Iraqi Governing Council, the other ministries, and the press. Allow Thamir to manage the ministry," I said.

Bahr Al Ulloum looked at me but did not say a word. He chose to ignore many of the recommendations that Phil and his replacement, Rob McKee, and I would make over the next several months.

Ghadhban remained in his position of chief executive officer, but Bahr Al Ulloum made every effort to limit his responsibilities. We would learn over time that Bahr Al Ulloum saw Ghadhban as a professional and political threat, nothing more. Bahr Al Ulloum could have really distinguished himself in the oil minister role. In the end, he failed to accomplish much of anything other than Chalabi's agenda.

The first key position that the new minister replaced was the director general of the State Oil Marketing Organization (SOMO), Mohammed Al Jibouri. Mohammed had been summoned to meet Dr. Ahmed Chalabi at the headquarters of the Iraqi National Congress (INC) at the Hunting Club in Baghdad shortly after Bahr Al Ulloum became the oil minister. He was instructed about how he would operate the SOMO organization going forward, especially the issuance of contracts. He was told to sell oil to Glencore, the international commodities company founded by Marc Rich in 1974. Al Jibouri objected. He would not allow sales to trading companies during his watch. Chalabi warned Al Jibouri. If he did not do as he was told, he would be replaced.[1]

Less than a week later, some negative reports about operations at SOMO were released to the press. The reports were false, incorrect, and misleading.

It did not matter; the threats made by Chalabi to Al Jibouri were being executed. Oil Minister Bahr Al Ulloum called in Al Jibouri to inform him that he was being replaced as the director general of SOMO.

Al Jibouri had been embarrassed by their methods. He took a leave of absence from the oil ministry. His reputation as a leader with untarnished integrity would remain intact and would serve him well in the near future. He returned to government service in May 2004 under the Allawi government as the minister of trade.

Ibrahim Bahr Al Ulloum removed Al Jibouri as the SOMO lead and replaced him with someone that followed instructions from Chalabi's team. SOMO started selling to Glencore and other traders, in spite of the previous CPA direction to sell only to large refinery companies. Why was it so important to Chalabi that Glencore become a customer of SOMO?

Marc Rich's company made billions of dollars in the Middle East oil trading business by moving Iranian crude oil through a secret pipeline in Israel. Glencore and Marc Rich had supplied most of Israel's oil needs since the 1970s.[2] With crude oil contracts from SOMO, Glencore was in a position to supply Iraqi oil to Israel.

CPA oil had just responded to an inquiry initiated by Wolfowitz at the end of August about no oil being sold to Israel.[3] The wheels were now in motion to change that. A more detailed discussion on Marc Rich, Glencore, and their movement of oil to Israel is covered in chapter 20.

While the Iraqis were changing out key positions within the Ministry of Oil, the CPA oil team was planning a major leadership change at the end of September. Phil Carroll returned to Iraq and prepared for his departure and to brief his replacement, Rob McKee.

Rob was the former executive vice president for Upstream at Conoco, the number-two position at Conoco below the CEO. Rob was a 1968 graduate of the Colorado School of Mines. He joined Conoco after graduation and stayed for thirty-six years, spending most of his time in upstream operations. He departed Conoco on April 1, 2003. Rob had no military experience, but his younger brother was a graduate of the US Naval Academy, a fact that he would soon let us know before the annual football classic in December.[4]

Rob was first approached in July to take Phil's position. He got a call from Dan Yergin at Cambridge Energy Research Associates (CERA). When Rob showed an inclination for the Iraq job, Phil Carroll contacted him. Paul Wolfowitz also contacted him from the Pentagon. He started the process of getting a security clearance and other approvals in July 2003.

Rob had known Phil from earlier days in the industry when Phil worked for Shell. They had met on several occasions and knew each other from their jobs in the oil patch. Rob also knew Don Evans, the commerce secretary in the Bush Administration, and he knew Vice President Dick Cheney from the vice president's time as Halliburton's CEO.

Rob arrived in Baghdad on September 20, 2003, for a quick turnover of the job. Phil left the next day, returning to Houston.

The end of September was also my demobilization date. My replacement was Pat MacArevey, who had been approved by the Pentagon and was ready to mobilize, but a late family issue prevented him from coming. He was a West Point graduate who also worked with me for several years at Mobil Oil, departing after Exxon purchased Mobil. Pat attended West Point after me but was well known by many of the younger general officers and colonels in Iraq.

MacArevey would eventually play an important role in the fielding of mine-resistant ambush protected (MRAPs) vehicles in 2007 as the vice president for government relations at International Truck. MacArevey's West Point classmate was Lieutenant General Frank Helmick, whose life was saved in Mosul in late 2007 when a vehicle-borne improvised explosive device (VBIED) exploded next to Frank's MRAP. Helmick walked away from his destroyed MRAP, which had been recently delivered to Iraq by his former classmate, Pat MacArevey.

Since Pat was a late cancelation and Rob was new to the job, I extended my tour at least until the end of 2003. I left Baghdad before September 20 for a break and returned during the first week of October, continuing in the same role under Rob that I played under Phil.

The rest of the CPA oil team included Terry Adams (United Kingdom), Brigadier General Bob Crear (Corps of Engineers), Gary Holcomb (DOE), Colonel Thomas O'Donnell (US Army), and Commander Kevin Ross (USNR).

Terry Adams joined the team in August. He was a BP upstream vice president with more than thirty years of experience. Terry had been the general manager of the Baku-Ceyhan pipeline through Turkey as an executive with BP. He was under contract with the UK government and stayed with the team for about four months before being replaced by Bob Morgan (United Kingdom) at the end of 2003.

Rob McKee saw his goals for the following six months as being in three primary tranches: strategic planning, refined products delivery, and oil ministry organization.

Strategic Plan

Rob focused our efforts for the oil sector on increasing crude oil production. The target level of 2.5 million barrels per day was reached early in the first quarter of 2004. However, attacks on pipelines in the north prevented sustaining that production number.

Attacks on pipelines were problematic, but that was not our biggest problem. The CPA budget director, retired Admiral David Oliver, had refused to fully fund the Task Force RIO program that had been developed in June 2003. He provided only about 60 percent of the original $1 billion target.

Oliver had committed to funding the RIO oil sector program in the summer of 2003 but backed off of that commitment in October. This reversal had a significant impact on oil production in the south.

His explanation for rejecting a funding request for Task Force RIO in October was that he refused to further fund the single-source KBR contract in the oil sector. The KBR contract had received a lot of publicity in the press and especially from Democratic congresspeople on Capitol Hill. Vice President Cheney had been the CEO of KBR and Democratic members of Congress were scrutinizing all expenditures on that contract. Unfortunately, there was no replacement to do oil reconstruction when RIO shut down.

The Oliver decision resulted in most of the 224 projects identified in June 2003 not being completed. CPA had agreed to complete these projects as part of the US government commitment to Iraq of repairing and restoring to prewar level. In fact, very few of these projects were completed. Most were started and stopped at less than 50 percent completion. Task Force RIO issued a "stop work" order to KBR in February 2004 as all funds dried up. The commitment made to the Iraqis in June 2003 that was based on a policy decision made by the president's cabinet was never kept.

Oliver left Iraq at the end of 2003 and left government service. He made two critical decisions during his time as budget director that severely hampered the reconstruction of the Iraqi oil sector: the summer payroll system and his funding reversal on Task Force RIO.

Two years later, the senior oil advisors reflected on their time in Iraq during 2003. We were doing a *Lessons Learned* report for the US Army. Phil Carroll, Rob McKee, and I agreed that Oliver's two critical decisions were the biggest impediments to completing our oil mission in 2003; the second place went to the insurgency and security.

Eliminating the capability of using KBR less than halfway through the oil reconstruction program was critical. It was a significant change from our prewar planning. The transition from the KBR single-source contract to the new Project and Contracting Office (PCO) system was a disaster.[5]

Why was it so bad? The US Congress established the Iraq Relief and Reconstruction Fund (IRRF) on November 6, 2003. It allocated $18.4 billion to rebuild Iraq's infrastructure. The oil sector was allocated $1.7 billion.[6]

The PCO organization administered the IRRF projects. PCO did not exist in November 2003; it had to be established. RIO started demobilizing in November 2003 after Oliver refused to fund their active projects. All of their

US-funded projects stopped work in February 2004. PCO did not start work on oil projects until the second half of 2004.

It became obvious to Rob McKee in late 2003 that a quick recovery of oil production could not be sustained without investing in downhole oil field work, drilling new wells, or work-over of existing wells. This was a change to our original plan. Up until October, the CPA oil focus was only on surface projects. The one exception was the repair of the seven wells damaged during the first month because of the well fires.

Rob and I attended a meeting with KBR and Task Force RIO at the Palestine Hotel in downtown Baghdad in October 2003. The engineers for KBR explained the huge benefits that could be achieved in the southern oil fields by implementing a well work-over program. KBR had recently completed work on the seven damaged wells in Rumaila oil field. On average, the wells were producing barely a thousand barrels a day before April 2003. After the KBR work, each of the seven wells was producing more than five thousand barrels per day.

Rob supported this effort to work-over several wells in the south to sustain the target oil-production levels. His support of this program in 2003 laid the groundwork for the eventual oil well work-over project completed in 2006 and 2007, three years later. Why did it take so long to execute such a high-priority project?

As previously mentioned, Congress appropriated the IRRF for the reconstruction of Iraq in November 2003. The oil sector was allocated about $1.7 billion of the total for three years, a small fraction of what was needed for the oil sector. Saudi Arabia was spending $20 billion per year at the time in their oil sector.

IRRF was managed under PCO, headed by retired Admiral David Nash. Nash needed to build the PCO organization from nothing; it did not exist prior to the appropriation. PCO contracted with Foster Wheeler Energy LTD for the program management of the oil sector. Parsons Iraq Joint Venture (PIJV) and KBR were awarded the two prime contracts to accomplish all engineering, procurement, and construction in the oil sector. The KBR team under the PCO contract was a totally different group from those on the Task Force RIO project. They were mobilized under a different management team and kept all of their work separate.

KBR-RIO was told by the Corps of Engineers to stop work in February 2004 when all funds were drained on their single-source contract. The PCO organization was just getting started at that time and would not be ready to accomplish any work in the oil sector until the second half of 2004. Rob's well work-over project was eventually awarded to PIJV in 2006 and work started late that year.

The oil well work-over project would eventually cost about $95 million to the US government. It involved well work-overs for eighty oil wells in the Rumaila and West Qurna oil fields. The benefits from the increased production out of those eighty wells were close to $20 billion over the following three years (2007, 2008, and 2009) in increased oil revenue for Iraq.

Refined Products

Shortages of refined products continued during the winter months of 2003 and 2004. Increased demand for gasoline and diesel products accounted for much of the shortages. For example, gasoline demand increased from about fifteen million liters per day in early 2003 to about twenty-five million liters per day in 2004, a 67 percent increase in demand. The domestic refineries could not keep up with that increase.

The higher demand was due primarily to a couple of factors. The first was the increased imports of new vehicles, both automobiles and trucks. The other contributing factor was the significant subsidy provided to the Iraqi population for refined products. For example, gasoline sold for about $.25 per gallon. Diesel sold for about $.20 per gallon. These low prices provided a huge incentive for smugglers to export fuel to the surrounding countries, where they could get market prices.

The eventual increase in Iraqi prices to approach international market prices started in 2004 and gradually increased during 2005. The subsidies continued but were not nearly as much as they were in 2003.

KBR was the primary importer of petroleum products and this was executed under the Task Force RIO contract. Imports of more than four hundred tanker trucks a day were common, importing from Turkey, Kuwait, and Jordan, and a little from Saudi Arabia. Tanker trucks from all over the Middle East and Asia started arriving to support this huge logistics operation. At its peak, we estimated that there were more than five thousand tanker trucks supporting the import program. Transitioning the administration of this volume from KBR to the State Oil Marketing Organization (SOMO) became a major challenge for the CPA oil team.

SOMO started importing tanker trucks from Kuwait, but their volumes were low and were not increasing. The reasons were numerous and SOMO eventually changed import strategies. They stopped tanker truck imports and went strictly to marine imports after the Basra ports started receiving oil ships.

We expedited project work at Khor Al Zubair's oil port for dedicated SOMO volume. This port had a pipeline to the Oil Products Pipeline Company depot in Basra, close to the refinery. The pipeline had previously been used to export diesel and heavy fuel oil. The ministry completed a project to reverse the direction of the pipeline to accept imports. This volume would eventually go north to Baghdad on the ten-inch product pipeline.

SOMO would also use the LPG terminal on the southern section of Khor Al Zubair for receiving products. This terminal had been originally constructed to export LPG and natural gasoline. Flows were reversed to import LPG and unleaded gasoline.

The KBR single-source contract again became a sensitive political issue back in the United States among Democratic congressmen. Product import responsibility could not successfully transition from KBR to SOMO in early 2004. SOMO was not ready to accept the responsibility for handling all the product imports. CPA had to come up with a plan to transition KBR to another US entity. That new entity had to handle the large volumes during the rest of 2004, until SOMO could successfully increase their volumes.

Managing imports of refined products transitioned from KBR to the Defense Fuels Supply Command (DFSC) in April 2004. KBR and the Corps of Engineers welcomed the opportunity to hand off this logistics mission. KBR did a nice job during the fuel crises in 2003, a mission that only seemed to get them a lot of criticism from members of Congress trying to politicize their efforts.

DFSC had the staff expertise and capacity to handle the contracts for supplying the large amounts of fuel. Imports continued coming across the borders via tanker trucks from Turkey, Kuwait, and Jordan. The significant change that occurred in 2004 under Rob McKee's leadership was the transition to importing substantial SOMO product volumes via small tanker ships and barges.

Attempts were made to import through port facilities on the Shatt Al Arab in Basra. This was necessary until the wrecks on the Khor Al Zubair were cleared. Abu Flus and Muftiyah Ports on the Shatt Al Arab, as well as Umm Qasr Port, imported small amounts of fuel from small ships or barges and loaded directly onto tanker trucks. This was a dangerous way to load gasoline. A small fire occurred at Umm Qasr that fortunately was contained; only the tanker truck was destroyed. The Umm Qasr port management refused to offload fuel after that accident. The risks were deemed too great. Umm Qasr was Iraq's primary dry goods port facility.

The Liquid Gas Wharf was converted from exporting condensate and LPG. It became an import facility in 2004. Pipelines ran from that wharf to storage facilities where trucks could be loaded. Pipelines running north to Baghdad from the storage facilities were eventually repaired and upgraded. The majority of the product imports quickly transitioned to the KAZ facilities during 2004. Pipeline and pumping improvements increased the capacity of this system. Most of Iraq's product imports during 2015 used this system.

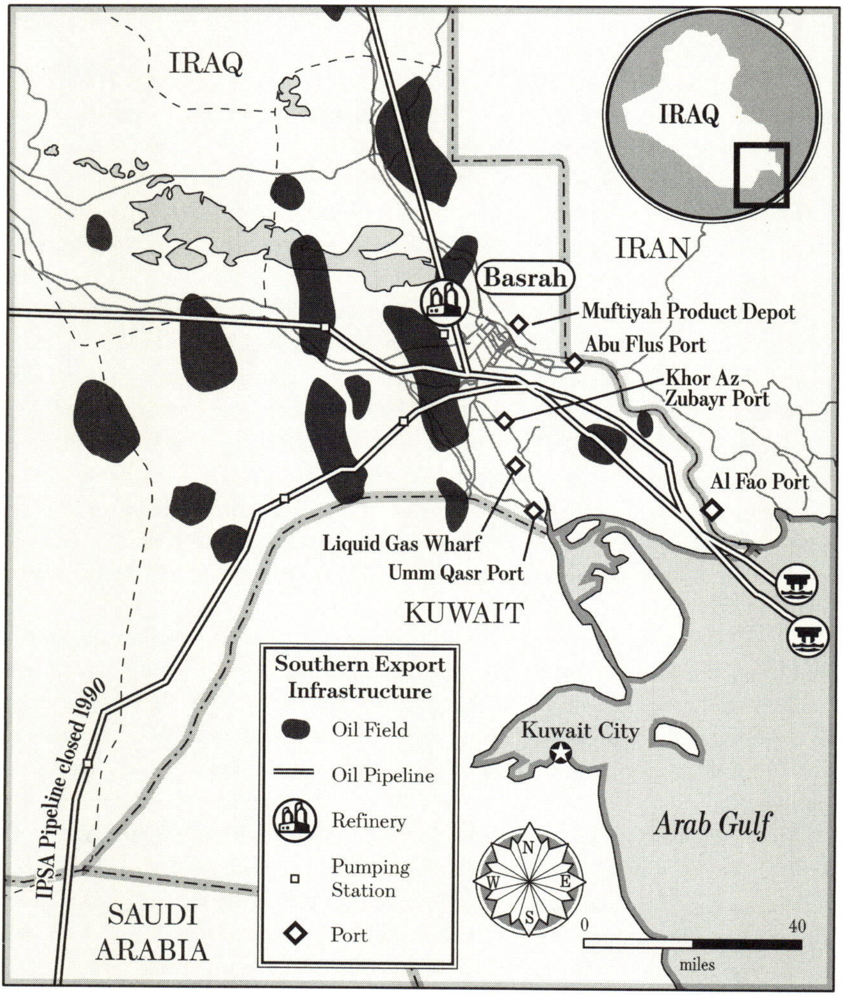

Southern Export Infrastructure as of 2003.

Organization

Rob McKee also focused on the need to change the organizational structure. Iraq had moved away from a national oil company structure during the late 1980s, placing the operator and regulator functions under the same person. The rest of the major oil exporting countries of the world maintained a separation between these two functions. For example, Saudi Arabia maintained the regulator function under their Ministry of Petroleum. The operator was the state-owned Aramco Oil Company.

One of the seven oil well fires in the Rumaila oil field, April 2003. Photo by author.

Gary Vogler and Phil Carroll in front of the Republican Palace after returning from a trip to the Ministry of Oil, May 2003. Photo courtesy of author.

The front of the Ministry of Oil headquarters on Route Irish in April 2003. Photo by author.

Rob McKee, Brigadier General Bob Crear, and Gary Vogler in the oil office at the palace, October 2003. Photo courtesy of author.

CPA oil team in front of the palace, March 2004. *Standing, left to right:* Gary Holcomb, Bob Morgan, Mary Salas, Mike Stinson, Rob McKee, Joel Berwick, Gary Vogler, and Colonel Tom O'Donnell. *Kneeling, left to right:* Colonel Lem Dubois, Kevin Ross, Captain Phil Noltemeyer, and Basile Qozi. Photo courtesy of DOD Photo Services.

A pond of crude oil forms next to the export pipeline near Basra after an attack on the export pipelines, June 2004. Photo by author.

Farewell meeting with Minister of Oil Thamir Ghadhban in his office on July 17, 2004, at the end of my first tour. Photo courtesy of author.

Thamir Ghadhban, Ambassador Robin Raphel, and Gary Vogler meeting in the PMAC office, October 2008. Photo courtesy of author.

Kevin Ross at the embassy helipad departing for a break after I returned to Baghdad to relieve him, February 2010. Kevin and I did back-to-back tours during our last two years in Iraq. Photo by author.

The Bayji Refinery in September 2011. The photo was taken as our Blackhawk helicopter took off after my last visit to the refinery. Photo by author.

Photo of the Mosul Dam taken from our Blackhawk helicopter during a flyover in September 2011. Photo by author.

The Al Basra oil terminal loading four very large crude carriers; each ship carries 2 million barrels. The gray ship in the background is a US Navy cruiser providing security. Photo in public domain.

A pipe-laying barge is installing pipe from shore in a canal that had been dredged the month before our flyover in September 2011. The pipe is part of the export project that became operational in March 2012. Photo by author.

Smoke billows from a gas-oil separation plant in the Rumaila oil field during our flyover in September 2011. Photo by author.

Three rigs await drilling from one of the international oil companies working in southern Iraq, September 2011. Photo by author.

Two drilling pads with active rigs creating new oil wells in the Rumaila oil field, September 2011. Photo by author.

Rob pushed for a study funded by USAID and contracted to BearingPoint that did an extensive review of all the major oil-exporting countries and how they were organized. BearingPoint engaged talented energy experts such as Amy Myers Jaffe from the Rice University Baker Institute and others. The report provided excellent detail for each of the seven countries studied: Kuwait, Saudi Arabia, Iran, Abu Dhabi, Venezuela, Indonesia, and Malaysia.

They started their study in November 2003 and a final report was issued on January 23, 2004. One thing that the report intentionally left out was a formal recommendation for Iraq to follow. The US government avoided making such recommendations for the Iraqi oil sector strictly due to political reasons. Recommendations really did not need to be stated. It became obvious that Iraq was the only significant player in the global oil industry that did not maintain a separation between the regulator and the operator. A structure with a separate Iraqi National Oil Company (INOC) similar to the organization in the 1980s was suitable. The oil ministry had accomplished their own study in the mid-1990s identifying the need to return to a national oil company.

We found a talented lawyer to help us with the organizational legal issues. Robert Maguire arrived in February 2004 and was assigned to the CPA Ministry of Trade advisory group. Robert started helping the CPA oil group immediately and we eventually poached most of his time from Trade.

Robert graduated with law degrees from St John's University, George Washington, and Georgetown. He had extensive experience in international law and worked in Moscow during the late 1990s. As a reserve marine corps colonel and former lawyer in the US Department of Energy, Maguire blended in well with our team. He would eventually extend his time in Baghdad to February 2006 before returning to the US State Department. He was appointed deputy general counsel and finally acting general counsel for the Department of the Air Force. He was one of the few Bush appointees under Secretary Gates who remained at the Pentagon during the Obama Administration.

Maguire's legal research revealed that the oil ministry had informally changed their organizational structure in the 1980s, but the Iraqi government did not pass any laws to eliminate the national oil company. The legislative foundation for a national oil company still legally existed. The existing oil minister could direct that the old structure be reestablished; no formal action by Bremer or a governing council was required.

Rob McKee tried to get Minister Ibrahim Bahr Al Ulloum to move forward with the organization change. Rob and Ambassador Bremer gave the minister substantial incentives, in the form of funding oil field projects from the CPA budget. They wanted Bahr Al Ulloum to move quickly toward the new structure. Bahr Al Ulloum ignored the recommendations.

A national oil company with its own board of directors of qualified professionals was no guarantee that the organization would not be politicized, but it would make it more difficult. Creating a professional oil company with separation from the politicians would have been good for the oil sector, but it would have taken power from Bahr Al Ulloum. It did not happen during his watch.

Reestablishing the Iraq National Oil Company (INOC) was part of Thamir Ghadhban's agenda during his short time as oil minister from May 2004 until early 2005. Unfortunately, security problems and product shortages stayed at the top of his priorities. He would later regret not addressing the organizational structure issue during his watch.

It became too easy for the oil ministry to eventually be politicized a few years later. The Maliki government of 2006 appointed a politician as oil minister and the political appointments went deep into the organization by 2012. Political affiliation rather than performance became the only qualifier for appointment to important leadership positions.

The most memorable occasion that Rob and I shared with soldiers occurred at the Thanksgiving dinner in 2003. A person from Bremer's front office walked into our oil office the day before Thanksgiving. He was trying to organize a bus for the next day to take several CPA folks to the airport for Thanksgiving with the troops. We felt like it would be a great way to spend Thanksgiving. Little did we know that it would become a Thanksgiving that we would never forget!

The bus pulled into an airport parking area close to the 1st Armored Division Headquarters, where a huge tent had recently been installed for Thanksgiving dinner. We entered the tent and saw tables, serving areas, and about six hundred soldiers from the 1st Armored and 82nd Airborne Divisions. We found a table of about ten soldiers and they made room for us two old guys to join them.

The organizers of the dinner said that food would be served after formal speeches were made, so Rob and I spent time getting to know our soldiers. We took time to get several pictures and talked for more than an hour. Everyone started to get hungry. Finally, the formal speeches started.

Lieutenant General Ricardo Sanchez gave a short talk and introduced Ambassador Bremer, who came to the podium. Bremer pulled out a piece of paper and said that he had a Thanksgiving message from the president. Then Bremer said that perhaps someone more senior should read it. President Bush emerged from behind the curtain to a crowd that could not believe their eyes. The place went wild. It was one of those moments in your life that sends chills up your back. You wonder if this is a dream or real.

Dr. Condoleezza Rice and Rob McKee during President Bush's memorable Baghdad visit, Thanksgiving 2003. Photo by author.

The troops rushed forward to greet him, and President Bush moved among them. He got behind the serving tables to fill soldier's plates with Thanksgiving turkey. Rob and I did not even try to get close to him. It was a day for the troops.

Condi Rice accompanied the president for that memorable meal. Rob and I approached her for a picture: "Ms. Rice, I am Gary Vogler, one of the CPA oil guys." She recognized my name and said that she knew Rob and me. Rob had met her before coming to Baghdad a few months earlier. One of Rob's memories of Condi Rice was that she, in his recollection, was the only Washington politician who ever asked him what she could do to enable him to do his job more effectively.

Everyone remained inside the large tent until the roar of loud jet engines flew over us. It was Air Force One taking off and heading back to the States. It will never be lost on either of us how the president stayed in that tent until every soldier got a photo with him, a hug, or a handshake. It was a wonderful demonstration of leadership.

The news of the president's visit spread through the press quickly. I phoned my family upon returning to the palace later that evening. They were getting ready to spend Thanksgiving Day with friends in our neighborhood. My wife saw the news reports and assumed that I was there. I had told her that we were doing something special for Thanksgiving, but we had no idea just how special.

Rob McKee reflected on his Iraq tour during my interview with him in Houston in March 2015. Rob gained a lot of respect for the professional Iraqis at the oil ministry. "All were patriots that did great work under very difficult circumstances," he told me.

Several senior American leaders also earned Rob's respect. "Ambassador Bremer had the hardest job in the world during 2003 and 2004. Generals Petraeus, Helmick, Odierno, Dempsey, and Crear provided great leadership, as observed from our oil patch foxhole," he shared.

Rob's biggest disappointment was the way our soldiers' efforts were being criticized in the press during his time in Iraq. He shared a story about recognizing one of the soldiers at Minute Maid Park in Houston not long after his return to Houston. The soldier also recognized Rob from his time in Baghdad. The soldier told him, "I thought we were doing a great job in Iraq until I got home and read our newspapers over here." Both lamented about how it seemed that the press had turned against the Iraq War in late 2003 and had used their stage for political criticism of the administration, as opposed to factually reporting the news.

This fact became apparent to many of us in the fall of 2003. I approached one of the reporters from the *Wall Street Journal*. His articles were becoming somewhat negative, while others seemed downright hostile. The reporter asked me if I watched Dan Senor's daily press brief.

Dan was Bremer's spokesperson and he held a daily press briefing at 1600 hours. The reporter said that Dan was spinning the news so much that the press felt obligated to spin in the opposite direction to keep the balance.

Dan was doing a lot of spinning, but it still looked like the press more than offset his optimistic spin. Unfortunately, this problem was much bigger than Ambassador Bremer's press spokesperson. It would become an even bigger issue upon my return to Iraq in 2006.

CHAPTER 13

CPA Closes Out

June 2004

Gary, Bob's gone, we lost Bob Morgan.
—Joel Berwick

The Bush Administration made the decision to transition from CPA to an interim Iraqi government no later than June 30, 2004. Ambassador Bremer had mentioned to Rob McKee and other senior ministry advisors in early 2004 that they could not leave Baghdad before July until properly relieved. In Rob's case, there was no one yet identified to replace him. He had been talking to an old Conoco friend, Mike Stinson. The two had discussed finding people from the US electric power industry to help in the Iraqi electricity sector. When it became obvious that the Pentagon could not identify an oil advisor replacement for Rob, he immediately asked his old friend if he would be interested in coming to Baghdad as the senior oil advisor.[1]

Mike Stinson was a 1966 graduate of Texas Tech University in engineering with an MBA from Arizona State University. Mike started working at Conoco in 1967 and left on March 1, 2003. He had thirty-six years with Conoco in several executive positions, including vice president of government relations in Washington, DC. Mike and Rob had worked together many years in the upstream business segment at Conoco.[2]

Mike arrived in Baghdad during February 2004, a week before Rob departed. February proved to be one of the calmest months in the last ten months. It would be the calmest in Iraq for a few years. This provided an opportunity for Rob, Mike, and I to accomplish a proper transition for Mike. The calm would be short-lived!

My final tour was supposed to end in December 2003, but there was no replacement. The proposed candidate was a West Point graduate who had worked in the Corps of Engineers while on active duty. He had spent many years with Mobil Oil and had constructed oil terminals in Africa and Asia. He had the exact skillsets for what we needed. Rob McKee showed me the rejection email from Mike Makovsky. Makovsky wrote that I was trying to do a favor for a friend. We got a chuckle over his email. The arrogance of someone who had never worked in the oil industry and refused to come to

Iraq was amazing. He was now making a judgment call on what we needed in the Iraqi oil sector.

As it turned out, things worked in my favor. Military mobilization orders for me were delivered. They ordered me to report to Kuwait in early 2004. The army reserve personnel office in St. Louis advised me to ignore the orders as long as I remained in Iraq as a civilian, or until reaching my mandatory military retirement date in July 2004. My last months would be spent as a US government civilian appointee in Baghdad rather than as an army officer in Kuwait.

We received a new industry executive in April. Norm Szydlowski had just retired from ChevronTexaco. Norm was the first executive recruited while still an active employee of an oil company. Norm had been vice president of Chevron's refining sector until accepting the job in Baghdad. He was a welcome addition to the team and would stay in Baghdad until the end of September 2004.

Ambassador Ronald E. Neumann, Robert Maguire, and Gary Holcomb accompanied a group of Iraqis on a delegation to Turkey in March 2004. Their mission was to negotiate issues surrounding supply contracts that were being terminated on the basis of force majeure and the recovery of Iraqi funds from oil smuggled in violation of the UN sanctions regime prior to liberation. It was during these meetings that Maguire learned that oil production sharing agreements (PSAs) were being made with Turkish companies by Kurdish leaders without legal authority, and in violation of UN Security Council Resolutions.

The Turkish company, Genel Enerji, signed a PSA with the Kurdish leaders in January 2004. Maguire pointed out that, as an occupying authority, CPA had the international legal obligation under the Hague Convention of 1907 to protect Iraqi government property. Such contracts violated Iraqi law and were inconsistent with this CPA obligations.[3] Maguire was very familiar with Iraqi law. He had traced Iraq's modern laws from their first constitution in 1925 up to and including their latest laws. He obtained a complete set of the English-language translation of Iraq's current Civil Code from the Library of Congress before coming to Baghdad.[4]

Like Phil Carroll and Rob McKee before him, Mike Stinson was vigilant in guarding against any oil grab during the CPA era and would not tolerate such activity on his watch. These confidential contracts signed by the Kurds were an affront to the authority of CPA and the Ministry of Oil. With the departure of CPA a couple of months in the future and the transition back to full sovereignty, our response had to be an Iraqi response, something endorsed by a responsible Iraqi authority.

Stinson and Maguire wrote a press release titled "Press Statement on Policy for Agreements to Develop the Natural Resources of Iraq." Maguire staffed the press release with CPA lawyers from State, DOD, Australia, and the United Kingdom. All of these stakeholders agreed with the international legal obligation of an occupying power to protect Iraq's natural resources. It was submitted for Bremer's approval and was eventually released under the signature of Oil Minister Ibrahim Bahr Al Ulloum in April 2004.[5]

The policy essentially stated that ownership of Iraq's natural resources belonged to all the Iraqi people and that any agreement to develop these resources remained a sovereign right of the government of Iraq. Also, any and all payments for the discovery, development, production, and disposition of petroleum and natural gas must be deposited in the Development Fund for Iraq to benefit the entire people of Iraq. The policy went on to state that non-Iraqi companies that wished to be welcomed in the future should not enter into or pursue the implementation of any agreement with persons not empowered to represent the sovereign government of Iraq.[6]

It was a well-written public statement and put all international companies on notice that for any development contract to be legally binding, it must be signed with the central government and not with tribal leaders or local authorities. The ink was not even dry on that governmental declaration before Kurdish leaders signed their next contract. As Bremer and his CPA staff made their final departure from Baghdad at the end of June, Kurdish leaders were signing a PSA with a small Norwegian oil company called DNO.

DNO was a small oil company with little financial resources. It was ready to engage in high-risk deals that would gain them high rewards. They teamed up with former US Ambassador Peter Galbraith, son of the renowned economist John Kenneth Galbraith. The younger Galbraith saw an opportunity to use his notoriety, former ambassador title, and contacts to gain DNO an illegitimate PSA and provide himself with a financial windfall.[7]

The US military experienced forty-two fatalities during the month of June 2004.[8] The US military and the US government civilians were tirelessly working in Baghdad to stabilize the population, promote a new cohesive interim government of Iraq, and safeguard Iraq's resources. At the same time, former Ambassador Galbraith was working with the Kurdish leaders to sign one of the first of a series of contracts. The contracts have proved divisive, have never been recognized by the Iraqi government, have created confusion, and have subsequently prevented the legitimate development of Iraq's northern resources to the benefit of all the Iraqi people.

Galbraith took advantage of the announced end of CPA and the instability of the new Iraqi government. Galbraith's supporters argued that he was not just motivated by money but also championed Kurdish rights. Whatever his motivations, he pursued financial claims against DNO for several years. In 2011, it was reported that DNO had resolved disputes with Galbraith by paying damages and legal costs as high as $92 million.[9]

Baghdad was relatively calm in March 2004. Some American civilians were still leaving the Green Zone on occasion and testing the security of the city. A few of us decided that we would attend Iraq's first West Point Founders Day celebration at Camp Victory near the airport. Founders Day was celebrated by West Point graduates all over the world during a designated evening in March. Joe Adamczyk, class of 1972, had retired from the army the previous year and was working in a civilian capacity at the palace. Joe was the organizer and the master of ceremonies of this historic event.

The road between the Green Zone and the Baghdad Airport would become the most dangerous road in Iraq within a few weeks, but it was calm this night. Four of us took a black BMW and made the journey at high speed just to decrease any risk. It was a memorable night, with more than two hundred graduates dressed in desert camouflage uniform. The oldest graduates were in civilian attire.

Founders Day gave me the opportunity to see many old friends. One had recently arrived. He was the new chief of staff for the 1st Cavalry Division, Colonel Bill Troy. Bill was a 1975 West Point graduate who would retire as a lieutenant general in ten years.

The Abu Ghraib Iraqi detainee torture incident was reported in the press in April 2004. Pictures were displayed all over the press and on CBS's *60 Minutes* back in the States. This was significant. Looking at the pictures, I felt embarrassment, remorse, and anger all at once. As an army officer, I was embarrassed to see this activity occur within the army. I felt remorse for those Iraqis involved and I felt anger at anyone in the chain of command with any knowledge that this had been happening. Brigadier General Janis Karpinski had refused to help me during the summer of 2003 with getting an oil sector director general released from prison. It did not surprise me when her name surfaced during this incident.

The next morning, I arrived at Thamir Ghadhban's office to review several issues in the oil sector. Two other directors were in Ghadhban's office waiting for my arrival. Hazim Sultan, the director general of the Reservoir Management Directorate, and Nabil Lamosa, director general of the Planning Directorate, were present. The three gentlemen were sitting on the cushioned chairs away from Thamir's desk. The tone was somber. They all invited me to sit with them.

"Gary, we are not going to discuss oil business today. We need to discuss something more important," Ghadhban said. Hazim and Nabil looked at me

West Point Founders Day dinner at Camp Victory near Baghdad Airport. Colonel Bill Troy and Gary Vogler, March 2004. Photo courtesy of author.

with somber faces. Hazim Sultan was a Sunni from Mosul who was educated in the United Kingdom. Nabil was a Christian who had worked closely with Ghadhban in the Planning Directorate for a number of years. Both he and Ghadhban spent time in detention in the early 1990s for criticizing Saddam's government.

"Gary, what do you think about this incident at Abu Ghraib?" Ghadhban asked.

"When I saw those pictures I immediately felt embarrassed. The three of you know me. I am a reserve army officer who spent almost eight years on active duty and then four years before that at West Point. I know of no army officer who would condone such activity. It is reprehensible," I said.

Hazim Sultan was always a cheerful person, but he was somber on this day. He mentioned that we came to Iraq and removed Saddam because he was a brutal dictator who tortured Iraqis. He proceeded to mention that now the Americans were guilty of the same type of torture. He wanted to know how we could justify these actions.

"I do not justify this. I will not justify this. Let me ask you, what do the three of you think the US should do to show that America does not condone this and are embarrassed about this?"

The three gentlemen looked at one another. I sensed that they felt like they had chewed on me long enough. It was time to discuss how to proceed.

Ghadhban looked at me and said, "Gary, four things need to happen. First, the US must publicly recognize the scandal and show remorse for their actions. Second, the hearings for this must be public. There must be full transparency around this incident. Third, identify all of the Iraqis that have been damaged by this incident and compensate all of those that have been hurt. Fourth, hold those people accountable that had some responsibility for this activity and it must be people higher than privates and sergeants."

Thamir's instructions were quite good. I took notes and put everything in writing and shared the document with Mike Stinson and several others within CPA. Most important, a copy was sent to my US senator from Virginia, Senator John Warner. Warner led the congressional hearings on this incident. His office acknowledged receipt.

The first three elements of Ghadhban's instructions were all met to some degree. The fourth element could be debated as to whether it was met, at least in a timely manner. Brigadier General Janis Karpinski was eventually demoted to the rank of colonel. She was responsible for all prisoner of war camps in Iraq. Lieutenant General Ricardo Sanchez was denied promotion to the four-star rank. He would retire as a three-star general.

Secretary Rumsfeld offered to immediately resign over the scandal, but the president refused to accept his resignation. The Abu Ghraib scandal became a cause for the insurgents. Attacks on US forces increased substantially after the scandal became public. New life emerged in the resistance. Military fatalities in the first quarter of 2004 were 119; they would more than double during the second quarter, to 257 killed in action.[10]

Ambassador Bremer and his staff started preparing for the transition to the first post-Saddam Iraqi government in early 2004. Identifying Iraqis to fill key positions in the future Allawi government occupied much of Bremer's time. The selection of oil minister was the focus for Mike Stinson and me.

Bremer called for a meeting. He mentioned that he needed to nominate a minister of oil. He knew the importance of such a selection and he had been disappointed with the performance of the current oil minister, Ibrahim Bahr Al Ulloum. Bremer was not familiar with any other competent technocrats within the oil ministry. Bremer asked Mike Stinson and me to nominate names for the oil minister position after CPA.

Mike and I looked at each other and both responded that we felt Thamir Ghadhban should be the oil minister.

Bremer said that he wanted to agree with both of us. He was impressed with Thamir when he had previously held that position, but there was a report that came in from the intelligence group that was very troubling for Bremer. He told us that he could not nominate Ghadhban based on that report.

A month prior, I had been sitting in the CPA oil office late in the afternoon by myself. A gentleman from the intelligence cell on the second floor walked into the office and asked if I knew a Mr. Thamir Ghadhban.

"Yes, I certainly do. I've been working with him for almost a year, quite closely," I said.

"We got this report in on him that is not good." The gentleman walked across the room and stood next to me. He read the report and described someone other than Thamir Ghadhban. When I heard the accusations my immediate thoughts were, "What a bunch of crap."

"Who is putting out those lies?" I asked.

"I am not allowed to tell you the source of this information," the gentleman replied.

"Well, tell your source that he is full of crap. Better yet, introduce him to me and I will tell him to his face that he is full of crap," I said.

The gentleman sensed my irritation and pointed to the lower section of his document labeled "Source of Intelligence." I read, in the block, "Iraqi National Congress" (INC). "Why is our Defense Intelligence cell still giving credibility to Chalabi and the INC after they had proven so unreliable?" I asked. He did not answer my question and it remains unanswered today.

When Bremer said that he had read an intelligence report on Ghadhban, I immediately thought of my meeting with the gentleman from the intelligence cell. I looked at Bremer and asked, "Sir, did you see the source of that information about Ghadhban?"

Ambassador Bremer replied that he did not know the source of the intelligence report.

"Sir, I saw the same report, and I saw the source. The source was the INC," I said.

Ambassador Bremer looked at me, his eyes lit up, and he said that he appreciated that piece of information. He agreed to support the nomination of Ghadhban.

Ambassador Bremer and Lakhdar Brahimi accepted Ghadhban's nomination. Brahimi was the undersecretary general of the United Nations and assigned to assist with the transition to an Iraqi government. It was Brahimi who interviewed and appointed Thamir Ghadhban as the oil minister in the Allawi government in May 2004. Dr. Ahmed Chalabi's INC was trying to discredit Ghadhban. The INC did not want him back in the key role of oil minister, a person they could not control.

• • •

It was early afternoon and hot on May 25, 2004. I was checking through the security point at the Baghdad Civic Center. There was a loud explosion that shook the earth where I was standing. The US soldier next to me and I hit the ground. We laid there flat on the ground, wondering if it was the only explosion we would get. After about two minutes, we stood up. "That was a close one," I said.

The soldier looked at me and shook his head in agreement. "I agree it seemed like it was just on the other side of the Green Zone wall next to us," he said. I would learn later that it was about a hundred meters down the road on the other side of the wall.

I finished my appointment at the Civic Center and was on my way back to the palace when my phone rang. It was Gary Holcomb at the oil office in the palace. "Gary, you need to get back as soon as possible. Joel cannot get into the trailer," he said.

The connection was poor, but I could tell that Holcomb was upset and that Joel needed my help. I immediately phoned Joel Berwick to see what the problem was. "Joel, what do you need?" I asked.

"Gary, Bob's gone, we lost Bob Morgan," Joel said.

I immediately drove to the palace parking lot and rushed to my trailer in the back of the palace. Joel was sitting on the steps with his head looking down between his legs. His shirt had bloodstains all over it.

Joel Berwick had arrived at the beginning of 2004. He was a Department of Energy employee assigned to a government office in Colorado. Joel was my roommate in our trailer. He was a sharp engineer who had been assigned three or four projects. One of those projects was assisting Bob Morgan with training the Iraqis at the Ministry of Oil on governance issues.

Bob Morgan was a retired BP employee who had left BP in 2001 after more than thirty years with the company. He joined the CPA oil group in February, filling the UK slot that had been vacated by Terry Adams a couple of months before. Bob had just extended for two weeks (for the second time) in order to complete the training that he was giving. He had received several policy and procedure manuals from BP and was using those manuals to train midlevel managers at the oil ministry on many different corporate governance functions. Bob was devoted to these issues and worked hard to instill modern policies into the Ministry of Oil and its operating units.

Morgan had been a well-loved member of the oil team. He had a great sense of humor and a disciplined approach to his daily schedule. After a long day of work, in the evening Bob would reach into the bottom drawer of his desk and pull out a bottle of the finest single-malt scotch. That was the signal for someone to get up and lock the door. Someone else would grab enough glasses and ice for whomever was left in the office. What a pleasant way to end a long busy day!

Bob Morgan and one of his personal security detail members were killed that afternoon from the explosion that forced me to the ground at the Civic Center. Bob had been directed to use the security firm Control Risk Group. All of the UK citizens assigned to CPA had to use Control Risk whenever leaving the Green Zone. They traveled in low-profile vehicles rather than the obvious convoys used by the Americans in CPA.

Joel was in the trailing Control Risk Group vehicle behind Bob when the vehicle exploded. It was engulfed in flames when Joel's vehicle pulled up adjacent to the burning vehicle. There had been two security personnel in the front of the vehicle and Bob was sitting in the back. One of the security guards rolled out of the vehicle and was quickly pulled into the backseat with Joel. That is when Joel was splattered with blood all over his shirt. That security person was rushed to the hospital inside the Green Zone and he survived. Bob Morgan and Mark Carmen from Control Risk Group did not survive.

Investigations indicated that the attackers had used a sticky bomb. It was the first attack in Iraq utilizing such a weapon. It had been placed on the back of the vehicle next to the fuel tank. When or how it was placed was anyone's guess. The most important message we heard in the oil sector was that it was a sticky bomb and our assumption was that Bob was targeted and that we were next.

Strict procedures were put in place after that for everyone to inspect underneath all vehicles before driving away, whether in the palace parking lot or at the Ministry of Oil. We all took this seriously.

Most of us previously refused to carry a weapon. The incident with Bob Morgan changed my attitude. A couple of others on the oil team also changed their minds. Colonel Tom O'Donnell had maintained a few extra Glock 9 mm pistols in the oil office safe that belonged to Task Force Shield. I carried one of those pistols in the bottom of a brief folder whenever I left the Green Zone during my last two months in Baghdad.

Mike Stinson did a superb job leading the oil group. Mike gave a wonderful eulogy during the memorial services held in the palace chapel in honor of Bob. There were several hundred people crowded into the chapel that day to remember Bob Morgan and Mark Carmen. Father Ken Beale, chaplain to CPA, presided at the memorial service.

Mike rallied the team to continue on with our work. Bob Morgan's desk was next to mine in the office. I found myself staring over at his desk several times over the following couple of months before my departure. It was the gloomiest period of my seventy-five months in Iraq.

Oil reconstruction slowed to a snail's pace in 2004 after the CPA budget director stopped the funding for oil reconstruction. The Task Force RIO organization started to demobilize soon after the stop-work order had been issued in February 2004. Colonel Lem DuBose had replaced Brigadier General Bob Crear in November 2003. Lieutenant Colonel Mark Schneider, USAR, replaced Colonel Lem Dubose in May 2004.

RIO was busy performing an administrative closeout of the incomplete oil projects under the KBR single-source contract. This contract had a ceiling of $7 billion, an estimate derived from the "worst-case scenario" of damage to the oil sector during our prewar planning. The actual reconstruction expenditures on this contract would amount to less than 10 percent, under $700 million. Many of the personnel from both KBR and the Corps of Engineers supporting Task Force RIO were leaving the country.

One exception was an active pipeline project in the north to transport oil from Kirkuk to Bayji. It included several water crossings that the Iraqis could not handle. This project was unlike previous RIO reconstruction projects. It was being funded with Iraqi oil revenues rather than US government funds.

The Iraqis had started constructing a forty-inch crude oil pipeline from Kirkuk to Bayji before we arrived in 2003. They were having significant problems with canal crossings and had requested Task Force RIO support. The Tigris River crossing was the widest water crossing for this pipeline and would become the most problematic.

Many of the oil pipelines between Kirkuk and Bayji crossed the Tigris River at the Al Fatha Bridge, just east of Bayji. Our military intelligence folks did not know this before the war. When the military developed the bombing target list and included the bridge as a target, it was not known that many important oil pipelines would also be destroyed.

Reconstructing the bridge was not a high-priority transportation project for CPA, but reconstructing the pipelines across the Tigris became a high-priority project for the oil sector. CPA refused to assign US government funds under the KBR single-source contract, but Iraqi oil revenues were deemed an acceptable way to fund the project. The follow-up contracts under PCO would not be available for another six to twelve months. The oil ministry considered this pipeline project too important to wait for US government funds.

The oil ministry was keen to use directional drilling under the riverbed for replacing the pipelines. Directional drilling is defined as the practice of controlling the direction and deviation of a wellbore to a predetermined underground target or location.[11] The oil ministry did not have access to directional drilling equipment before the war and was eager to use such equipment; this is how most pipeline river crossings are currently done throughout the rest of the world. The Iraqis wanted to use directional drilling for better security of the pipelines. They also felt that directional drilling would be the most effective and least expensive way to traverse the river.

The Corps of Engineers took soil core samples of the riverbed and identified that there would be some engineering challenges due to the significant amount of cobblestone in the area where the drilling would take place. It was felt that the driller could move the drilling site along the riverbank if the cobblestone in one area was too severe. The Iraqis and RIO accepted the risk and KBR brought in a subcontractor with good experience drilling under the Houston ship channel and elsewhere. Unfortunately, the timing of attempting such an engineering challenge could not have been worse.

The directional drilling equipment was mobilized and delivered to the site in early 2004, just before the Abu Ghraib incident hit the press. The spring of 2004 saw an uptick in hostilities against all US installations. Just after the Abu Ghraib prison incident, the US military also initiated combat operations in Fallujah and Najef areas.

The Al Fatha crossing project received more attacks than any other project site. It was an easy target in a remote site with few people and less than a dozen civilian guards carrying rifles. Mortar rounds started hitting the site almost daily.

With the directional drilling operators spending a lot of their time seeking safety in shelters, the work moved slowly. The contract was for a specified number of workdays at a daily rate. A day sitting in a protective shelter avoiding mortar rounds was also a paid workday.

Mike Stinson and Nabil Lamossa from the oil ministry visited the site shortly after Mike took over as the senior advisor in April. A couple of Blackhawk helicopters transported the two of them to visit the work site. Mike saw the location and observed the difficult security situation that the KBR subcontractor was working under. "Everyone would stop work and haul ass when the rounds started coming in," Mike said. They would slowly return to work if no rounds came in after awhile. It only got worse after their visit.[12]

Thamir Ghadhban approached me in early May to get an update on the project. He had been hearing that there were problems and he wanted to know the status. Our request for a project status report took several weeks to receive. It was finally delivered in June: "We are 25 percent complete and out of money." The report was a surprise to all of us.

It was not that our status requests were being ignored; in fact, our request was made each day during the morning oil meetings. We heard detailed reports every day on the security situation and why a status report from the work site was not available. The RIO project manager was physically located in Kirkuk and she could not visit the site because either RIO security deemed the Al Fatha work site as too dangerous or the route to the site was too hazardous.

After hearing the status, CPA oil refused to seek more funding for the project. It was stopped and the contractor demobilized. The security situation would not allow for the project to continue. Mike Stinson called a meeting of the RIO and KBR managers to seek answers, a postmortem report of the project.

Mike, Norm Szydlowski, and I attended from the CPA oil group. Our conclusions pointed to the poor security situation at Al Fatha as a reason for the failure of the project. It also had severe technical issues due to the presence of cobblestone. The poor security situation denied any opportunity for a possible workaround solution to the cobblestone problem.

The contractor could only successfully pull the smaller twenty-six-inch pipeline. The larger, forty-inch pipelines were too large for the cobblestone and the larger pipe made up most of the scope of work.

It would take another two years for the US forces to successfully construct a pipeline crossing at Al Fatha, which they did in mid-2006. Parsons Iraq Joint Venture (PIJV) under contract from the Gulf Region Division of the US Army Corps of Engineers would dredge a trench in the riverbed and install several pipes in the trench. This was attempted only after the US forces cleared and secured the surrounding area.

The Al Fatha pipeline project would highlight the difficulties of successfully completing any oil project in Iraq when the security deteriorated to a point that made the costs prohibitive. The Special Inspector General for Iraq (SIGIR) decided to audit this specific project and issue a report in early 2006, long after the departure of everyone who participated in the postmortem audit conducted by Mike Stinson. SIGIR's audit challenged the competency of the Corps of Engineers. SIGIR stated that the reason for the failed project was technical and essentially ignored the security problems, sarcastically calling it the "Al Fatha Pipe Dream."

The SIGIR audit would place a black mark on the reputation of the oil reconstruction work of the US forces. The *New York Times* ran an article on this project in April 2006 based on the SIGIR audit, using SIGIR's incompetency theme and ignoring the issue of poor security. The Iraqi press had similar articles that ridiculed the US forces for their Al Fatha pipeline-crossing project. The Iraqi press and the oil ministry would start labeling our military engineers the "Corpse of Engineers," citing the publicly distributed SIGIR audit from January 2006 as their source for ridicule.

The Al Fatha audit report and all of SIGIR's audits were published on an open website available for anyone to use or abuse. The press loved the open source made available to them; "it makes our job much easier," a reporter commented to me. Audits are accomplished in any organization to identify problems. Publishing these problems on an open website for anyone to review had consequences.

The SIGIR audit report on Al Fatha would have unintended consequences the year following its release, 2007. The open-source audit would eventually cost the Iraqis about $12 billion in lost export revenues from 2008 through 2010. How so?

South Oil Company (SOC) wanted to transfer close to $100 million from their annual capital budget to the US government in 2007 after the US government oil reconstruction funds were gone. The SOC funds were to increase the scope of the highly successful well work-over project administered by the US Army Corps of Engineers in the southern oil fields.

Production and exports were declining in the south in 2007. The Iraqis needed help to reverse the trend. The well work-over project was increasing production by three thousand to five thousand barrels per day for each oil well completed. This project could provide the needed help.

The US leadership in 2007, General Petraeus and Ambassador Crocker, welcomed the opportunity for the Army Corps of Engineers to help with such an important project. The oil minister was onboard, but the prime minister's office refused to approve, citing the poor reputation of the US Army Corps of Engineers in the Iraqi press.

The Iraqi press kept sourcing the SIGIR website to highlight their argument that the US Army Corps of Engineers was not competent. It was not the opinion of the Iraqi press that our engineers were incompetent but that of our own auditors, SIGIR. I was advised that the weak Maliki government could not take the criticism that would follow by giving Iraqi funds to the US Army "to squander, as reported in the SIGIR audit of Al Fatha."

The eventual decline in southern exports cost Iraq more than $12 billion over the next few years. The decline started in 2007 and was not reversed until the international oil companies started managing the oil fields in 2010.

Attacks on pipelines decreased each month through April 2004, but that trend quickly reversed after April. The most damaging attack was on the export pipelines in the south that decreased or eliminated exports for about two weeks. This was a critical time just before the transition to the new Allawi government.

Two large pipelines supplied crude oil from the southern oil fields to the Al Fao export terminal. Both pipelines were attacked within a few hours at two different locations. One pipeline was successfully attacked at a location about five miles west of Basra. The other was hit south of Basra where it traversed the mud flats.

These southern attacks were coordinated and required inside knowledge for identifying the specific locations for the attack. Both pipelines were buried and were not the only large pipelines in the area. However, they were the only active export pipelines.

I had planned a trip to Basra and heard of the attacks just before my departure from Baghdad. After landing at the Basra Airport late in the day, I linked up with the Task Force Shield team located in the area. We visited the attack sites as the sun came up the next morning.

The first site visited was the forty-two-inch pipeline site that took place on the mudflats south of Basra. An explosive charge had been successfully placed where the pipeline passed next to the service road. The SOC repair team was on-site and had been working all night using a lighting system and generators. They were surprised to see me pull up but quickly welcomed me. They had successfully cut out the damaged pipe and had the new pipe-joint hanging from a crane, ready to be lowered into position once the backflows of oil stopped. It would take several more days for the pipe to drain.

The second attack site was west of Basra and would be the more difficult to repair. It was the forty-eight-inch pipeline buried a couple meters below the surface. The SOC repair crews had not yet started to repair this pipe before our arrival. A large black oil pond was growing while we were there. Oil gurgled up from several feet below us.

My final stop was late in the day to the SOC headquarters to see the director general, Jabbar Al Luaibi. I had not seen Jabbar in several months and wanted to bid him farewell before my departure the following month. He greeted me in his office.

He mentioned that SOC was putting most of its efforts toward the site that should be easier to repair, the forty-two-inch pipe. It was easier to handle and was not buried as deep as the forty-eight-inch pipeline. He fired the senior security manager and the pipeline repair manager that I had seen earlier in the morning. I had heard about the security manager removal. Word of this spread quickly through the Task Force Shield channels. Jabbar did not waste any time. He had a reputation for being a decisive leader. His new appointment in the security job had a reputation for being a professional and was a former general officer in the Iraqi army. Firing the pipeline repair manager was a surprise. He seemed to be making good progress when we visited him at sunrise.

Jabbar explained that the repair manager refused to push his people to fix the damaged pipeline. He defended his decision by saying that Iraq could not afford to wait until everything is perfectly safe. Iraq had to take more risks than companies do in the West.

Jabbar was a demanding leader. He put a lot of pressure on subordinates to perform. His leadership style served him and Iraq very well that first year and would continue for the next five years, until his retirement.

Repairing pipelines after attacks became a high priority in 2004 and remained a high priority for years after. More Iraqi pipelines were attacked in one month than most oil companies experienced in leaks over decades. Over three thousand attacks on oil pipelines were recorded through May 2008 on the Product Pipeline Company's seventy-two hundred kilometers of pipelines.[13]

CPA oil and the oil ministry could not repair pipelines fast enough. The ministry's Product Pipeline Company started refusing to repair certain important pipelines that were in dangerous areas, such as the eighteen-inch crude pipeline for the Doura Refinery. It kept getting attacked in the triangle of death, southwest of Baghdad. The CPA oil team needed to come up with a better plan, a different way to approach our problem.

We interviewed a new infantry officer that had just arrived from the Pentagon and had been assigned to Task Force Shield, Lieutenant Colonel Tony Cusimano. Tony was a career regular army officer who wanted a challenge. He did not want to spend the war in a headquarters office. We could not have asked for a better officer to take on our pipeline repair problem.

Tony met with Saleh Aziz, the director general of the Products Pipeline Company, to better understand which repair jobs Saleh would handle and which repair jobs were too dangerous, and why Saleh considered them too dangerous. As the security situation deteriorated during 2004, the areas considered too dangerous increased. Tony studied the threat in each of these areas based on information received from Saleh and especially from any intelligence available from the US forces.

Cusimano linked up with the 1st Cavalry Division, which had taken over Baghdad security responsibilities from the 1st Armored Division earlier in 2004. The division leadership was eager to help since they knew that quickly repairing pipelines meant more oil products in Baghdad. The 1st Cavalry assigned a platoon to escort Tony's repair team to some of the most dangerous areas outside of Baghdad.

Tony was initially assigned a pipeline repair team of American civilians under a Corps of Engineers pipeline repair contract. Unfortunately, these Americans refused to go to the areas where Tony needed them to work. It was too dangerous. He had to find other workers, so he contracted with a local Iraqi company that agreed to work with him and the 1st Cavalry.

Tony's repair team was a success. Their missions were frequently sent to the triangle of death to repair the crude oil pipeline supplying the Doura Refinery, or they went north to repair pipeline breaks to the crude oil pipeline between Kirkuk and Bayji. Those pipelines supplied crude oil for export in the north. The northern export pipeline route between Kirkuk and Bayji had become very dangerous in 2004.

Saleh Aziz depended on Tony's team to handle the ever-expanding list of jobs in the "too-dangerous" category. Lieutenant Colonel Cusimano got his wish. He spent little time in an office during his tour in Iraq.

The 1st Cavalry honored Tony before his departure from Iraq in late 2004 with the highly coveted cavalry spurs. He was inducted into the Order of the Spur, a cavalry tradition normally reserved for members of a cavalry unit serving in combat. His missions went into some of the most dangerous areas and received more hostile fire than any other military unit at the time. Not a single soldier was seriously injured during any of his missions.

The soldiers felt that Tony had "good karma" that protected them. They requested parts of his dog tag necklace to keep his good karma with them in Iraq. Tony joked that his dog tag necklace had become a choker by the end of his tour.

Kevin Ross departed Iraq in mid-July 2004 and I left a week later. Mike Stinson left in August. Even though hostilities were on the rise, I felt good about the oil sector. Strong leadership was in charge. Good leadership was important and CPA left Iraq with the best leaders for their oil ministry. Thamir Ghadhban was oil minister. Strong leaders surrounded him in all of his key management positions. The directors general throughout the ministry were appointed based on performance and not politics. This would change significantly over the next decade.

Meanwhile, the fatalities to US forces continued to increase. There were 849 fatalities in 2004, in addition to the 486 fatalities in 2003.

CHAPTER 14

A Break between Tours

2004 to 2006

Gary, get your suitcase packed.
—Oil Minister Thamir Al Ghadhban

Anyone returning from Iraq after military or civilian tours experienced a period of decompression. I was no different. My first tour lasted seventeen months. I had spent more time in Iraq than any other American. I was physically and mentally exhausted. Civilians do not normally fire weapons, but we got fired at. Many times our responsibilities resulted in the loss of lives if we did not execute effectively. This is a heavy burden for anyone.

There must have been some type of exit interview by someone at the Pentagon. My only recollection of communications with the Pentagon upon my return home was my final leave and earnings pay stub that arrived in the mail. I recall looking at it and just thinking that was the end. My government service was finished.

Unfortunately, it was not that easy. Like many of my colleagues, I became an Iraqi news junkie. My days started with reading everything in the news that I could find on Iraq. There were not many Iraq books written yet, but I read everything that was on Amazon. It got to the point where Amazon would notify me whenever they started selling a new book on Iraq.

Gold's Gym became home for two hours every day. My physical conditioning improved quickly while my psychological adaptation adjusted slowly.

It is amusing to look back at those days now. I saw the movie *American Sniper* in 2014 and found myself laughing when others in the theater were wondering what was so funny. One scene showed Chris Kyle gazing at the blank screen of a television set in his Texas living room. That seemed to be my favorite position in late 2004 while my brain relocated back to Iraq. It made no difference whether the television was turned on or tuned to golf; my mind was in Iraq, reliving my tour.

Various groups that were doing studies or were holding conferences on Iraq contacted me. Most were very gratifying. The one that left the biggest impression on me was a study conducted by the Dupuy Institute in Annandale, Virginia. This was a small group of PhDs who had received a contract

Norm Szydlowski, Hussain al Fatal, Gary Vogler, and Dathar al Khashab at oil ministry for appointment of new Oil Minister Thamir Ghadhban on June 3, 2004. Photo courtesy of author.

from the US Army. They studied insurgencies over the last eighty years to understand the current insurgency faced in Iraq.

Dupuy brought me in toward the end of the study to help them confirm their previous findings. They were almost complete with their work when they shared all their findings and conclusions with me. The two conclusions that I kept remembering over the following years were that the insurgency that started in the summer of 2003 would probably last ten years or more and that the United States would incur five thousand to ten thousand killed.

This study was completed at the end of 2004 but did not receive much publicity. The Dupuy Institute and Christopher Lawrence finally published a book in 2015 to document their work.

The loss of life was not limited to coalition military and civilians. Iraqis incurred fatalities and life-changing injuries on a much greater scale than we did. *Business Insider* reported that more than 134,000 Iraqis were killed between the invasion in 2003 and the end of 2014.[1] There were a few incidents involving employees of the Ministry of Oil during my tour in 2003 and 2004, but none affected me as much as the assassination of Hussain Al Fatal in October 2004.

Al Fatal was competent and honest. He was appointed to the position of director general for the Products Distribution Company in the fall of 2003 after the previous director general was attacked. Hussain prided himself on his work weeding out the corruption that he saw taking place within the Distribution Company.

Hussain and I worked closely together. He and I attended several meetings with US military officers that were trying to ensure that their areas of operation were supplied with product. Hussain impressed all those who met him, including the future chairman of the Joint Chiefs General Marty Dempsey.

During my last meeting with Hussain in July 2004, he spoke of the many nefarious characters that approached him to engage in corruption. He laughed as he described the way he cursed these characters and chased them out of his office. Unfortunately, they caught up to him in October.

Similar to the attack on his predecessor, the attackers went after him close to his home. They observed how Hussain walked his young daughter to the bus stop to send her off to school for the day. One morning in October 2004, after waiting for the bus to depart, the attackers drove by and sprayed bullets, killing him. It was a very difficult loss for my Iraqi friends and me.

A retired army major from the Pentagon PCO oil organization phoned me in 2005 and requested my help to do an infrastructure security assessment for them. He said that I was the perfect fit for the job, if I was interested. I agreed. A week later, he phoned me back to let me know that Mike Makovsky had rejected me. No reason given. He did not have any other candidates identified and asked me to recommend someone. Lieutenant Colonel Tony Cusimano was working on the army staff at the Pentagon. Tony would have been my first recruit if I had taken the job.

Cusimano got the assignment to lead the effort. He contacted me several times during the project for my comments and recommendations. We exchanged emails and spoke on the phone many times. My presence in their planning area was banned. I was blacklisted.

I did not know then just why the neocons blacklisted me. My blacklisting became obvious to several of my oil colleagues, but no one knew why. I inquired with Makovsky in December 2005 when he and I attended the same meeting at the State Department. He confirmed that I was blacklisted but refused to share why.

The Corps of Engineers contacted me in 2005 and requested that I lead a *Lessons Learned* review on the Iraqi oil sector. The Bush Administration was beating the war drums to move into Iran and the Corps of Engineers did not want to be unprepared in case they needed an oil contingency plan.

It was an enjoyable project because it brought together my colleagues from Iraq. Input was received from retired Colonel Tom O'Donnell on Task Force Shield, previous managers with KBR such as Stoney Cox, and even a report from an oil ministry director general. I received written input from Phil Carroll, Rob McKee, and Mike Stinson. The four of us met in Houston late in 2005 to finalize the report. Although the report was unclassified, it was not published publicly until 2016. Some of our key findings and recommendations in the *Lessons Learned* report are mentioned throughout this book.

An Iraqi delegation visited Washington in December 2005. It included Thamir Ghadhban, who was being considered for the oil minister position in the new Iraqi government. We held a reception for him during his first evening in Washington and invited all of the military and civilians who had worked with the Ministry of Oil during their Iraq tour. We had about thirty people in attendance from around Washington, DC.

While standing around in groups before the dinner, Ghadhban said to me, "Gary, there is a good chance that I will be appointed oil minister in a few months. Can I count on you to come back to Iraq and help me?"

"Thamir, you know that I would love to work with you again, but Barb would have to support such a decision and she has been resolute against me volunteering for another tour," I said.

"Give me ten minutes," Ghadhban replied and immediately excused himself from our group and joined a group of folks in the corner of the room where Barb was located. I noticed from thirty feet away that he was focusing all his comments toward Barb. He returned to our group as we were taking our seats for dinner: "Gary, get your suitcase packed."

Later that evening during our ride home, I asked my wife what Thamir Ghadhban had discussed with her before the dinner. Her only response was, "You cannot go back to Iraq until after our daughter's wedding in September 2006." He did it. Ghadhban had amazing interpersonal communication skills!

The year passed by quickly and I returned to Baghdad in early December 2006. Ghadhban did not get the oil minister position, but I had been given my green light to return and I wanted to. My seventeen-month tour in 2003 and 2004 was the most rewarding work in my life. I thoroughly enjoyed working with the Iraqi oil professionals and wanted to continue. Plus, a few members of the US government Baghdad oil team requested that I return.

I applied for a job in the embassy oil group but was rejected. The rejection confused me. The folks that contacted me in early 2006 requested that I apply for the senior oil advisor position. There were two other applicants competing against me, but neither approached my experience level. My suspicion was that my rejection was because the neoconservatives had blacklisted me. Two government employees later confirmed my suspicions. One worked in the Baghdad embassy oil group and the other worked at the Pentagon.

With the US government path blocked, I accepted an opportunity as a contractor with Foster Wheeler Energy Limited (FWEL). FWEL had the program management contract for oil reconstruction with the Gulf Region Division of the Corps of Engineers. Returning as a contractor offered me a more generous leave package that enabled me to sustain a long tour for the following five years. It offered other benefits as well.

Robert Maguire returned from Baghdad in early 2006 and worked at the State Department for several months. Robert had the most recent Iraq oil experience, so I sought out and valued his counsel. He encouraged me to accept the FWEL contractor position, a technical job requiring engineering skills that I had not used in several years.

Robert recognized that being a contractor had advantages because it would provide a barrier between the internal politics of the Baghdad embassy and me. He also let me know that I would be considered a CPA person and CPA had become the target of blame for everything that went wrong. His guidance was to avoid the embassy and spend all of my time in the field.

Robert advised me to find a project that the Iraqis wanted and needed but that was not on the embassy's radar screen. That was great advice and became my challenge. It took a few months and the arrival of a new commanding general, but a project was found that would eventually be labeled by the oil minister as the most important Iraqi infrastructure project in decades.

Several things surprised me upon my return to Baghdad. First was the poor security between the airport and the Green Zone. The security was bad in 2004 but not nearly as bad as in 2006. I took a C-130 from Amman, Jordan, that landed at the Baghdad Airport late in the afternoon the first week in December. No daylight ground movements were authorized that day because of security problems. We waited until after dark to catch a UK forces Puma helicopter for a ride to the Green Zone.

My bags were packed behind and underneath my seat on the side of the helicopter. There was barely enough room for six passengers and their baggage. The door of the Puma was open, the toes of my boots were outside of the aircraft, and the gunner was next to me firing flares every couple of minutes as a defense against heat-seeking missiles.

It was a cold night and we flew just a couple of hundred feet above Baghdad at high speed, doing banking and turning maneuvers the entire trip. It was a ten-minute flight that provided more excitement and adrenaline rushes than any Disneyworld ride. I missed the adrenaline rushes. It was good to be back!

Navy Commander Jim Sellner greeted me at the helipad next to the US embassy and took me to the two engineer compounds. Jim was the deputy director of the oil reconstruction group. He was a Navy Reserve Civil Engineer Corps officer who was a mining and geological engineer for the state of Minnesota when he was not in uniform.

The engineer compounds were located inside the Green Zone, a kilometer from the embassy. Our living quarters were adjacent to the 86th Combat Support Hospital, the military hospital compound. We lived in new two-story dormitory buildings with very small private rooms, a shower, a small refrigerator, and a television. The people that worked at the embassy still lived in the trailers that we used during the CPA era, so I felt that my accommodations were an upgrade. The dining facility was next to our dormitory and our offices were across the street on a separate compound.

Foster Wheeler had hired me as a subcontractor. Their senior manager in Baghdad was Chris Humphrey, a UK citizen who had worked on many projects with Foster Wheeler. He was a competent engineering manager with a solid background and understanding of industry management practices for large capital projects. Foster Wheeler would complete their contract and depart Iraq in the summer of 2007, but I would remain as a subcontractor for other companies for another five years.

Chris had discussed my hiring at length with the US government's director of oil projects, Captain Richard Fritzley, USNR. Fritzley reported to a Corps of Engineers senior executive civilian, Les Dixon, who was responsible for all reconstruction projects in Iraq. Les reported to Brigadier General Mike Walsh, commanding general of the Gulf Region Division (GRD), and Mike reported to General George Casey, who would eventually be replaced by General David Petraeus.

The GRD main office was located on Essayons Compound inside the Green Zone. GRD had three districts in Iraq: North, South, and Central. The Central offices were located adjacent to the Baghdad International Airport. The South offices were in Basra and the North offices were in Kirkuk. Most of the US government employees were military field grade officers or Department of the Army civilians. They administered the reconstruction contracts for all sectors: oil, electricity, and water. The oil contracts were with FWEL, Parsons Iraq Joint Venture (PIJV), and KBR.

Fritzley had extensive private sector experience in capital project management as the manager of project planning and controls at Boise Cascade, a major paper and packaging manufacturing company headquartered in Boise, Idaho. His private sector management skillsets were good to excellent and he had spent more than twenty-five years as a Civil Engineer Corps officer in the US Navy Reserve. Rich welcomed me to the oil team and quickly shared his plans for me.

Fritzley and his predecessor had significant disagreements with the embassy's senior oil advisor on several of the oil projects and wanted fresh eyes and an independent opinion for moving forward. He looked to me to provide that opinion.

Some historical context should be added here. Robert Maguire had been working at the embassy when John Sickman was hired to the senior oil advisor position. Robert shared with me the history of this particular hiring. Sickman had been hired for a budget analyst position but got two promotions before his plane landed in Baghdad at the end of 2005. His oil industry management experience was the weakest of any previous senior oil advisor. It included managing an offshore production platform in the North Sea for a small American upstream oil company. An electric power company later purchased his company and sent him to Peru to manage a gas pipeline.

The previous senior oil advisor, Bob Todor, and his deputy had resigned over the Christmas holidays of 2005 and did not return to Baghdad. It became very difficult for the State Department to recruit and retain competent oil industry professionals to work in Iraq during that period. Robert Maguire was working in the oil group when Sickman arrived at the beginning of 2006 and Robert departed in February.

Todor and all previous senior oil advisors had worked for large integrated international oil companies at the executive level and had extensive experience. Unfortunately, Sickman's experience was much more limited. Team building, international oil pricing, refinery operations, downstream oil blending, and major downstream maintenance best practices experience were not skillsets needed for a budget analyst. However, they were extremely important for the senior oil advisor position. Sickman was lacking in all these skillsets.

Fritzley would need the help of a much bigger gun than I had available. It would actually take the arrival of General Petraeus to make the needed changes.

The year 2006 had been a terrible year for Iraq. The dome of the Al-Askari Shia Shrine in Samarra had been destroyed in February 2006 and the security in the country had deteriorated since then. An emerging civil war was creating a very poor security situation.

The new government of Nouri Al Maliki had appointed Dr. Hussain al-Shahristani as the new oil minister in May. He had no experience in the oil industry and had never successfully directed any large organization in his life. His most noteworthy accomplishment was refusing Saddam when he was appointed to lead Saddam's nuclear program. Shahristani spent several years in prison for rejecting the nuclear job. Some Americans considered him another Nelson Mandela because of his long political imprisonment. He may have been a good politician and may have been an acceptable oil minister during normal times, but these were not normal times for Iraq. His appointment would prove to be an unfortunate choice for this period. Also, the oil advisory services provided to him by the US government during his first year were the weakest provided to any oil minister since the liberation.

My initial activities included a trip to Basra to meet the rest of the US oil team and review the major projects in progress. Foster Wheeler was the program manager and had several good engineers in Basra: Robert Carruthers, Scott Monson, Alan Wayne, and Greg Waner. We had two prime contractors still active in the Basra area doing the engineering, procurement, and construction (EPC) work: KBR and PIJV).

PIJV had started a project that proved to be one of the more successful in the southern oil fields, the well workover project. As I stated in a previous chapter, this project had its genesis when Rob McKee and I met with the RIO and KBR engineers in late 2003. At a cost of about $95 million, the project essentially paid for itself before the final well had been completed, about nine months after the first well. The scope of work included eighty wells in both the Rumaila and the West Qurna oil fields.

There were fifty wells in the supergiant West Qurna oil field that had been partially drilled before 2003. PIJV perforated and stimulated each well. The South Oil Company installed the flow lines to the gas-oil separation plants (GOSP) and each well started contributing three thousand to five thousand barrels per day to exports.

There were thirty wells in another supergiant oil field, Rumaila, that were shut because the water content in the production had increased to a level that could not be satisfactorily processed by the surface equipment. Those wells had to be plugged and reperforated in another production zone. They immediately added to the exports at the same volume as the West Qurna wells. All eighty wells added enough exports for the following few years to increase Iraq's revenues by about $20 billion.

Lieutenant Colonel Steve Loken, USAF, was Fritzley's deputy in Basra and he directed the effort in Basra from the US government side. Lieutenant Colonel Harry Callicotte, USAR, was assigned to work on this specific project. Callicotte had been in Iraq for more than two years working on several oil-related issues in CPA, the embassy, and the Gulf Region Division of the Corps of Engineers. He had been in the class of 1979 at West Point and had spent much of his civilian career in the oil patch. This would be his final project before ending his service in Iraq.

Loken directed all the contractors involved: PIJV, Weatherford International, and Foster Wheeler. Besides the US contractors, the South Oil Company provided many of their Iraqi engineers and workers. This team effort worked to keep the costs low and provide training to the Iraqis. It was a huge accomplishment and would earn both Loken and Callicotte much-deserved commendations for their work.

PIJV started a major project on the Al Basra oil terminal (ABOT) during my visit. ABOT was Iraq's primary export terminal in the south. This project would be completed in mid-2007 and represented the last major oil reconstruction project fully funded by the US government. New meters, new wiring, control equipment, and firefighting equipment were installed as part of the ABOT project.

KBR was preparing for phase two of a major maintenance project for the South Gas Company. Phase one had been successfully completed a couple of months before my visit and phase two of the $145 million project would start in early 2007.

I met all the managers for both PIJV and KBR during my three-week visit. All were eager to complete their commitments under their US government contract and exit the country. Both companies had worked in the north and the south and were doing great work for the Iraqi oil sector, but they saw the security situation growing worse and did not see any opportunities for them. There was little interest in staying any longer than necessary.

The security situation in Basra was better than in Baghdad, but it was still bad. The UK forces had moved out of the city and headquartered at the Basra Airport.

Foster Wheeler had engineers colocated at both the KBR and PIJV compounds. It was recommended that I stay at the KBR compound, where there were more than adequate facilities. It was adjacent to the UK forces perimeter at the airport. Mortar rounds hit us just about every evening. My caravan was fortified with sandbags around the outside and there were concrete shelters just outside my door.

The normal procedure was to grab a personal helmet and flak jacket and move as quickly as possible to the shelter. We would remain in the shelter and text our status to the operations center over our cell phones. The operations center would then give the "all clear" and we would move back to our caravans.

During my second week in Basra, I experienced my worst attack. It was after 11 PM and I was sound asleep in my bed.

An initial rocket landed close to our compound and the alarm was sounded to move to shelters. As I jumped out of bed and turned on the lights, a huge explosion hit.

Most rockets used by the Shia militias were the 125 mm Iranian-built rockets. Although these rockets were lethal, it normally took a direct hit to do serious damage. This particular rocket attack included the 240 mm model and these larger rockets were deadly. The sound was deafening and my caravan shook as if it was going to crumble. The insulation and dirt fell from my ceiling. Diving under my bed, I grabbed anything I could find to cover me.

There did not seem to be enough time to get to a shelter. After a minute of quiet, I quickly grabbed my helmet and flak jacket and moved to the concrete shelter to wait out the remainder of the attack.

When the "all clear" was finally given, I returned to my caravan. It was a mess. The next morning, I entered the engineering offices and they were also in disarray. A line of a dozen KBR Filipino office employees formed outside of the KBR administration office. Several employees resigned. They were paid quite well with the hazard premium versus what they were paid in their home country, but they refused to accept the risk anymore.

It was understandable. The attack the previous night had been intimidating. I was fortunate. My only injury was a rug burn I received on my leg when I lunged under my bed. No one was killed and there were no serious injuries, but I suspect there were many minor injuries like mine. I began to have second thoughts about the wisdom of my return to Iraq.

Upon returning to Baghdad, I made two disturbing reports: one to my Foster Wheeler supervisors and the second to Captain Fritzley.

Foster Wheeler had heard initial reports about the Basra rocket attack from their employees and wanted to hear directly from me. We held a conference call with the FWEL corporate security manager and I shared my Basra experience. This created a serious dilemma for him. No Western company ever wanted to be in a situation where employees or subcontractors were in harm's way. Had the situation become so serious that employees should be evacuated? This was being seriously considered.

My report to Captain Fritzley was more of a warning. KBR had been directed by the ministry and the embassy's senior oil advisor to modify all their major maintenance planning for the South Gas Company during the summer of 2006. The original plan had envisaged one phase with all work completed between September and November.

This plan had been in place for close to a year and all the engineers assigned to the project by KBR and Foster Wheeler had agreed to stay on the job until the end of 2006. The engineers were unwilling to extend their stay. They moved to their next jobs in other parts of the world in December. Many of the new jobs were with other companies.

The embassy's senior oil advisor and the new oil minister had insisted on making the last-minute change to split the project into a two-phase project rather than a one-phase one. Phase one of the project had been successfully completed, but all new engineers and managers had to be assigned to phase two, scheduled to start in January.

The decision made by the oil minister and the embassy senior oil advisor was contrary to the best practices being used in the industry and raised the risk of delays and higher costs to the overall project. Our practice at Mobil and ExxonMobil was to designate such major maintenance work as a project that engineers and managers were assigned during the planning phase and that they remained on until completion. No one was removed from the project unless they resigned from the company.

My warning was not a surprise to Fritzley. This was one of the major disagreements that Fritzley's predecessor, Navy Captain Mike Sherbak, had with the embassy senior oil advisor.

Unfortunately, my warning became reality a couple of months later. We had to move funds from other projects to cover the cost overruns. The oil minister became irritated when delays to the project started to cost Iraq additional LPG imports at a cost of tens of millions of dollars. Minister Shahristani wanted to blame KBR for these additional costs. His inexperience was the real problem.

The embassy oil group held morning meetings similar to those meetings we held during the CPA era. Rich Fritzley normally sent his deputy, Commander Jim Sellner, to attend these meetings. Sellner was in the field, so I volunteered to attend in his place. Gary Holcomb got me into the embassy for the next meeting. Holcomb was still a Department of Energy employee on loan to the embassy. He had remained in Baghdad since the CPA days.

Holcomb welcomed me at the embassy gate and got me through security. He warned me that the current oil group, especially the senior oil advisor, was much different than the oil professionals we had during CPA. I was not going to be dissuaded. There was too much at stake; the dysfunction caused by bad chemistry had to be fixed. Developing teamwork among employees was important.

The meeting surprised me. My expectations were low, but I was still surprised. The senior oil advisor, John Sickman, walked into the room of about ten people and started his meeting. Holcomb interrupted and introduced me. I thanked Gary and mentioned that I was looking forward to working with everyone. Sickman ignored my comments and me. He went on with his agenda. He finished his meeting and quickly walked out of the room. It was obvious to everyone in the room that I had been snubbed.

My resolve to fix this problem took a hit. How could I make a dent in this dysfunction if the lead oil person refused to talk to me? I was just a contractor. I had no real authority. My personal credibility and ability to sell ideas were useless if the oil lead refused to even talk to me.

I questioned my decision to return to Iraq. The situation in Iraq, especially the dysfunction in the oil sector, was much worse than I had thought it would be. The security situation had progressively gotten worse. Military casualties were higher than in 2004 and getting higher. I had almost become one of those casualties in Basra the previous week. That would alarm anyone. On top of that, the person who held the job that I had held in 2003 refused to talk to me. Why fight it? After being in Iraq for more than a month, it was obvious that I would not be able to contribute in a positive manner. It was time to leave. I returned to my desk and drafted a letter of resignation.

Word quickly spread throughout the offices; President Bush had just announced that General George Casey was going to be replaced by General David Petraeus. I last visited with Lieutenant General Petraeus in July 2004 just after he returned to Iraq in the job of training the new Iraqi army. He had impressed me with the work he did commanding the 101st Airborne Division in Mosul during 2003.

I sent him a short cordial email congratulating him on his latest promotion and new job. I mentioned my return to Iraq and that the situation was discouraging. He had a huge challenge in front of him.

A return email showed up in my inbox less than thirty minutes later. Petraeus sent me a detailed email full of questions; most were specific questions about the Bayji Refinery. He knew that Bayji was the largest supplier of products in the country and that it was the largest employer in the volatile Sunni area. It was critical for supplying essential services to the Iraqis and providing a boost in the Sunni economy.

There was not much information for him on the Bayji Refinery. It was not on the Gulf Region Division's radar screen and my limited information came from operations reports that the embassy oil group distributed. I gave him a good overview of some of their problems, and they were all serious problems. One positive fact that I shared with him was that the oil minister had just appointed a new refinery manager, Dr. Ali Obaidi, whom I had worked with in 2004. Ali was a good leader and would play a significant role over the next few years.

We exchanged more emails over the next couple of hours and Petraeus concluded by telling me that he would be in Baghdad in two weeks and expected me to fix all the Bayji problems before his arrival. His comment was made in jest. The problems were significant and would take many months to fix, but it was good to hear positive comments coming from someone who could fix them. His positive attitude was contagious! My letter of resignation was never submitted.

I reached out to John Sickman via email. Petraeus had a vision for the Bayji Refinery, a facility that everyone had forgotten about. With my refinery experience and understanding of the Bayji units, Sickman and I could work together on this one. John had never worked in a refinery. To my knowledge, none of his people had ever worked in a refinery. This was my opportunity to fix the dysfunction. Sickman responded to my emails but refused to meet with me. This was shared with Captain Fritzley and he took my request to his higher headquarters.

The embassy oil group reported to the Iraq Reconstruction Management Office (IRMO). Ambassador Joseph Saloom headed this organization and all civilians reported to him, including the senior oil advisor. Saloom had a military deputy who had several military personnel inside IRMO reporting to him. He was also responsible for coordinating IRMO activities with the military units, especially our engineering unit, the Gulf Region Division (GRD). The current military deputy was Major General Steve Abt, a reserve general officer whose civilian job was as a professor of engineering at Colorado State University.

Unknown to me, Abt had ordered Sickman to meet with me. I received an email from Sickman agreeing to a meeting at the embassy coffee shop in an hour.

Arriving at the coffee shop as soon as possible, Sickman and his lawyer were waiting to meet me. Yes, he brought his lawyer. To my surprise, it was an antagonistic reception—no friendly hello and no cordial greetings. Sickman told me that I had thirty minutes to state my business. The meeting only lasted ten minutes.

We were on two different wavelengths. I tried to get answers to General Petraeus's questions so that we could start a discussion and develop a plan to help the Iraqis fix their largest refinery. The lesson that was chiseled into my brain from my previous tour in Iraq was that shortages of products led to instability. Instability resulted in attacks and deaths of US soldiers. I explained this to both Sickman and his lawyer but got no reaction. John was more interested in government bureaucratic procedures concerning me as a contractor and his own job security. I never walked away from a meeting in my life. This was my first.

Things were going to change and change quickly with the arrival of General Petraeus and Ambassador Crocker. There was too much at stake. The new surge strategy was necessary to prevent a catastrophe.

General Petraeus established an Energy Fusion Cell (EFC) soon after he arrived in Baghdad. The EFC was directed by a British general officer and took the lead on electricity and oil issues. The EFC's oil director was a US Army colonel. I worked closely with him. The IRMO oil group was disbanded before the summer of 2007. A few members of the group joined the Embassy Econ Section and the others went home, including John Sickman.

CHAPTER 15

Energy Fusion Cell

Fixing the Oil Sector

It was about security, security, and security.
—General David Petraeus

General Petraeus arrived in Baghdad in early February 2007. His subordinate commanders started to reassess their areas in light of the new commander's intent. Brigadier General Mike Walsh, commander of the Gulf Region Division (GRD) of the Corps of Engineers, ordered a review of the major infrastructure project areas under his command: oil, electricity, and water.

One of the first issues that we addressed in the oil sector was identifying any single points of failure. What inherent risks in the oil sector could bring down the government of Iraq?

Our team conducted a review of the oil sector that focused on every asset under the Ministry of Oil: refineries, pipelines, gas plants, oil fields, depots, and export facilities. No single asset within the oil sector could affect the government of Iraq in a negative way more than the southern export facilities. These offshore loading terminals came under attack in early 2004 by suicide bombers piloting three speedboats. Fortunately, the attack was successfully repulsed. After that, coalition navy resources secured those offshore facilities.

The offshore oil terminal, Al Basra oil terminal (ABOT), handled 95 percent of Iraq's exports, and oil exports accounted for 95 percent of Iraq's revenues in 2007. The export channel in the north, the Iraq-Turkey (IT) export pipeline, was constantly under attack by insurgents. The IT pipeline seldom operated for more than a day or two before attacks shut it down. We believed the single point of failure for the Iraqi government in the oil sector was ABOT. It contributed export revenue of about $100 million per day by exporting daily volumes of about 1.6 million barrels. If this facility were destroyed, where would the money come from to address the shortfall?

The offshore facilities seemed to be secure from external risks; the coalition navy had resources on the terminal and patrolled the surrounding waters. The navy had taken over the helipad at the southern tip of the offshore terminal and installed sophisticated radar equipment; it also operated small drones to investigate any suspicious vessels in the distance. At least one large warship patrolled close to the terminal, providing significant intercept capability.

However, there were internal threats that had not been previously addressed. For example, the smaller and older offshore facility close to ABOT was called Khor Al Amaya oil terminal (KAAOT). It had incurred severe damage from an industrial accident in 2006. This accident caused an explosion that essentially put most of KAAOT out of operation. Repair work to restore the facility took several years.

The poor maintenance practices used by the Iraqis—welding in the presence of hydrocarbon—contributed to the industrial accident on KAAOT. Not only did the accident severely damage the terminal, but it also killed several welders.

Coalition navy personnel on ABOT observed similar maintenance practices on that terminal in 2007. An incident occurred in early 2007: Iraqi welders wanted to weld in an area on ABOT where oil had not been properly cleaned up. The US Navy personnel confronted the Iraqi supervisor and vacated that area of the platform before the welders could start working.

Operational problems also increased the risk of a catastrophe. One of our GRD contractors on ABOT witnessed an incident in early 2007. While observing the loading operations of a very large crude carrier (VLCC), our contractor heard a loud vibration noise and the platform started shaking. Recognizing the problem, our guy immediately ran to a location on the platform and grabbed a large hammer. He used the hammer to bang on a sixteen-inch relief valve. The noise and shaking continued until he got the relief valve to activate, sending pressurized oil to a surge tank.

The loading operation had experienced a pressure surge known as a fluid hammer when the Iraqi operators abruptly stopped the oil flow. This was not unlike a water hammer that sometimes occurs in a home when a faucet is quickly closed. The relief valve on ABOT was very old and poorly maintained. It did not activate when the pressure in the pipeline surged, and this created the hydraulic shock.

The oil flow on the platform during loading operations would fluctuate between forty thousand and eighty thousand barrels per hour through the forty-eight-inch pipelines. The Iraqis would sometimes have to suddenly shut down the oil flow. Such activity is normally acceptable if a proper operating relief valve can quickly divert flow to a surge tank on the loading platform. Unfortunately, the relief valves on ABOT were old and seldom operated properly. In this case, the force behind the high volume of crude oil going through a forty-eight-inch pipeline was great enough that it could have caused damage to the subsea pipelines or the terminal.

It did not take much discussion within the GRD oil group to agree that ABOT was our single point of failure. Once agreement was reached within GRD that the offshore oil terminal was a serious risk, our challenge was to identify how we could help the government of Iraq mitigate the risk.

GRD sent me to Basra in March 2007 to meet with the leadership of South Oil Company and discuss possible projects that could be executed by GRD. Our focus was primarily on the security of onshore pipeline systems, but the South Oil Company management was much more concerned about alternative export channels, a backup for ABOT.

We held a meeting during March at the South Oil Company Technical Department offices near the Zubair oil fields. It was a small but strategic meeting. Mr. Oday Al Quoraishi, the capital projects manager, attended from SOC. We discussed several possible onshore projects, none of which made any sense to the South Oil Company management. Oday shared the concerns of his senior management. He mentioned that his boss, Director General Jabbar Al Luaibi, was well aware of the high risk to their offshore export terminal and badly needed a backup southern export channel.

Oday mentioned that the Japanese had been working with the Ministry of Oil headquarters since 2005 on a project to install a new subsea pipeline and a single-point mooring, but that project appeared to be stopped. Only discussions and studies were taking place over the last couple of years. No procurement or contract work was progressing.

A single-point mooring is a buoy floating on the surface, anchored to the ocean floor by chains. A pipeline supplies oil to the buoy from shore, and flex hoses transfer the oil from the buoy to the ship. The technology for such loading facilities had improved over the last forty years and, in many cases, was considered the most cost-effective loading technology at that time.

Oday and I used empty soda cans left over from an earlier meeting to design a single-point mooring project in the northern Arabian Gulf. Little did we both realize that our design made with these cans would be the groundwork for what would become a multibillion-dollar project over the next five years. At this point, we agreed that the best way for GRD to assist with the backup export opportunity was to help progress some type of single-point mooring project with the Ministry of Oil.

Upon my return to Baghdad, a meeting was scheduled to visit the ministry. Specifically, we needed to talk to Deputy Minister Ahmed Al Shamma, who was responsible for coordinating with the Japanese on their export project in the south. Their project had been announced in 2005 as the Sealine Project with a scope of work to install one forty-eight-inch pipeline to ABOT and include the installation of a single-point mooring with the new pipeline.

The meeting with Deputy Minister Shamma went well. It was the first time that I had seen him since my return to Iraq. He welcomed me into his office and we discussed old times from 2004. We finally got around to discussing the intent of the visit. Rich Fritzley and I shared the desire of the US forces to utilize some funding for infrastructure security on projects to help South Oil Company and their export infrastructure. When it was mentioned that SOC was keen on the United States assisting with installing a single-point mooring, his eyes opened widely. He was interested.

The deputy oil minister quickly shared some of his concerns with the Japanese project. Bottom line, the Japanese were studying the export project but did not have any desire to start construction until the security situation in Iraq dramatically improved. Deputy Minister Shamma welcomed our offer to pursue a single-point mooring project with South Oil Company and the oil ministry.

Shamma had been a project engineer assigned to the ABOT construction project in 1975 and was well aware of the vulnerabilities of the old, decaying facility. He had personally managed a single-point mooring project for SOC in the early 1990s, but that project ended when the contractor went bankrupt. He mentioned that one of their buoys was still in dry dock somewhere in Southeast Asia.

Fritzley and I shared the results of our oil ministry meeting with the GRD leadership. We were directed to draft a formal letter from General Petraeus to Prime Minister Maliki identifying the export infrastructure in the south as a high risk to the government of Iraq and offering the assistance of the US forces to help develop a project for installing single-point moorings. The letter was approved, signed, and given to Prime Minister Maliki by General Petraeus in March 2007.

The details surrounding this project will be covered later in the book. Suffice it to say, Brigadier General Mike Walsh had scored an early win for GRD and his reputation with General Petraeus. Mike was a career officer in the Corps of Engineers with a lot of expertise in government funding for projects and capital project execution. He did wonderful work in leading the project execution for the US government projects in Iraq.

It became evident that General Petraeus needed something more from the oil and electricity sectors. He needed more than he was getting from the existing command structure. It was all about security, security, and security. His focus when it came to electricity and oil was on how the two sectors were interdependent and supported his primary mission of maintaining security and stability within the country. The organizational structure of GRD and IRMO worked well to manage the multibillion-dollar reconstruction efforts, but Petraeus needed something more responsive to his security needs.

The stories from those morning battle update assessment (BUA) meetings of early 2007 highlighted irritation from General Petraeus. He was not getting answers or was being told that it was someone else's responsibility when he inquired about operational issues within the energy sector. He needed a single point of contact for addressing his energy security needs.

General Petraeus looked for someone to lead the energy sector similar to the way Brigadier General Frank Helmick had led the energy sector in Mosul back in 2003 and 2004. Petraeus commanded the 101st Airborne Division in Mosul back then. He needed that one general officer with an organization that would be his principal advisor for any energy-related issues: electricity, oil, and the security of those sectors. He needed an Energy Fusion Cell.

The leader of this Fusion Cell had to have both good technical and good leadership skillsets in order to be effective with this new organization. The Baghdad bureaucracy would resist any threat to their perceived established boundaries, so the Fusion Cell leader must have the right personality and excellent communications skills.

Lieutenant General Graeme Lamb of the UK forces was General Petraeus's deputy. Lamb offered a solution for finding the "new General Frank Helmick"—a Brit. What a great idea! The American bureaucracy would be less inclined to perceive a threat from a Brit. A young brigadier general from the UK forces was scheduled for deployment to Baghdad in May. General Lamb had worked with this young general officer previously and was impressed with his problem-solving abilities. Plus, the new guy was an engineer officer. His name was Brigadier General Richard Cripwell.

Cripwell arrived in May 2007 for his six-month tour. General Lamb escorted him into his first BUA and then Cripwell accompanied Lamb into the follow-up meeting after the BUA known as the "small group session." This was the short meeting that followed the daily BUAs that consisted primarily of the more senior people in the command.

General Lamb introduced his new general: "CG, this is Cripwell. He is the genius I brought in to sort out the Energy Fusion Cell."

Petraeus looked at Cripwell, smiled, and said, "Welcome, you have twenty-four hours to fix it."

The reality was that Cripwell was given a "pass" for about five days, and then Petraeus started challenging him, asking him detailed questions about the energy sector that forced Cripwell to learn quickly. It was an on-the-job (OJT) training program with intensity—OJT on steroids.

Reflecting on those days eight years later, Cripwell highlighted his time under Petraeus and those many long morning meetings. "Petraeus would lean heavily on me and the team to impress upon his command, the hundreds that listened into the morning BUAs, that he placed a lot of emphasis on getting the energy sector moving in the right direction. The morning meeting participants included most of the field grade officers in the country, general officers, and senior commanders. Petraeus used those meetings in an effective way to communicate to his command."[1]

Other senior leaders under Petraeus shared Richard Cripwell's opinion. Major General Ken Hunzeker invited me for dinner and cigars one evening. Hunzeker was the senior advisor to the Ministry of Interior and commanding general of our mission to train, man, and equip the Iraqi police forces. He had been in the country for about ten months. We had known each other since he was a lieutenant colonel, several years before. Ken's compound was located about two hundred meters from our engineer compound inside the Green Zone. We met for dinner at his location and sat outside in the warm evening, drinking cokes and catching up.

Ken had worked for General Casey for about five months and now had been working for General Petraeus for the last few months. I asked him, "As one of the senior general officer leaders, how would you compare the two leadership styles: Casey and Petraeus?"

Hunzeker replied,

> Under General Casey, the morning meetings were used as a battlefield update assessment with most of the meeting giving updates and little to no time on the assessments side. General Casey did most of the talking, with very few questions. I don't ever remember him asking me a question in any briefing. Keep in mind, he had been here for more than a year when I arrived so he knew much more about the interior ministry and our mission than I knew at the time. Under General Petraeus, the format of the briefings stayed the same, but the discussion was all about doing an assessment, giving daily guidance, and asking for your opinion in open forum. General Petraeus used the meeting to set the operational pace. For example, rarely would a morning meeting go by without Petraeus asking me one or two specific, detailed questions about my organization and our mission and how it was going. The questions caused me to be a lot more prepared, and recognize that I was now part of the assessment process. I believe my organization worked much smarter than we were a few months ago because of these detailed questions, discussion and assessments.

The observations by Brigadier General Richard Cripwell and Major General Ken Hunzeker were shared by many of the senior leaders under Petraeus. Those of us that monitored the morning BUAs on the video teleconference admired his leadership style. There were critics, and their criticism often was in the form of "he micromanages"; some commented that Petraeus had an ego. From a businessman's perspective, I welcomed such micromanaging. The oil sector was dysfunctional. My experience at Mobil Oil and Exxon-Mobil taught me that the best leaders always got into the details when there were problems. Iraq had plenty of problems. Also, most successful civilian and military leaders had ego issues; it was part of the terrain.

Petraeus knew how to bring out the best in his sharp, talented subordinates. He knew how to challenge subordinates on a multitude of levels: intellectual, physical, emotional, and the like. I enjoyed watching or listening to those morning BUAs and often felt like I was observing one of the best military leaders our country had to offer. His intellectual ability to quickly grasp and retain complex electricity and oil issues set him apart from even our best general officers. It was an honor to be part of what would soon be recognized as a military success, the surge.

An Iraqi leader in the oil sector gave the best compliment that I ever heard anyone give Petraeus. The Iraqi's name was Riyadh Ghassab, and he held the position of director general in more oil ministry companies than any other Iraqi. The most difficult time of the surge was early summer 2007. We were losing more than a hundred soldiers a month and it looked like little progress was being made. Riyadh had retired from the oil ministry earlier in 2007 and was doing consulting work for me under GRD. In light of how bad the situation was in Iraq, I asked Riyadh what he would do if he were President George W. Bush? Riyadh said that he would appoint General Petraeus as the viceroy of Iraq for a minimum five-year term of office. He went on to say that Petraeus was the only leader in Iraq, American or Iraqi, looking out for the benefit of all Iraqis.

Richard Cripwell took his first week to observe and assess his new position and the challenges in front of him. A summary of his notes from that first week included several critical observations of his new organization: "dealing with tasks and not problems, no traction, we know a lot of people but none are linked up, responsibilities and authority not aligned, muddied priorities, no standard procedures, a focus on day to day issues, and a lot of duplication of effort."[2]

Cripwell saw his role as designing a framework and then trying to get the main players to buy in. He spoke to all of the oil and electricity players and asked a simple question, "Are you with me or not?" This was important because very few of the energy resources reported to him. Most fell under the embassy IRMO or Econ sections and several remained under GRD. One person simply replied to his question, "You don't write my report."[3]

The Energy Fusion Cell had no money and no authority but plenty of responsibility. Cripwell became the center of attention for all things related to energy in the eyes of the commanding general. Petraeus looked to Cripwell for the answers on energy. Over time, all the serious stakeholders saw Cripwell as the guidon bearer for the energy sector. When General Cripwell took a grilling during the morning BUAs, it was normally because Petraeus was communicating a point to his subordinate leaders.[4]

Cripwell's personality and communication skills were good-natured. Many agreed that he came across as the scrappy guy with high energy. At the same time he was bright, a quick study, and could articulate problems and solutions better than most. He became our scrappy guy in the Energy Fusion Cell. Richard won over almost everyone on our team, and the group came together to protect him as the months passed.

Cripwell built a bank of credibility by taking those morning BUA poundings. The division commanders out in the field respected him, similar to the way one respects the boxer who keeps getting pummeled by his opponent but bounces back and stays in the fight until he wins.

When he would approach the field commanders to get their support with electricity issues in their area of operation, the response at first would be "hey, I am fighting a war down here. I don't have time for electricity." It did not take long before his reputation with Petraeus became known throughout the command. He started getting much more priority with the field commanders. A good example of how Petraeus supported him was use of the CG's personal helicopters.

Petraeus demonstrated his early support of the Energy Fusion Cell by making an offer for Cripwell to use his personal helicopters to support the EFC's mission. Cripwell took advantage of the offer on several occasions.

Iraqis in the electricity ministry always knew within an hour when a major high-voltage distribution line was cut. Locating the downed pylon or power line in Iraq took an eyes-on approach. Someone had to physically go out and pinpoint the location of the outage. Richard would call Petraeus's executive officer, Colonel Pete Mansoor, to get the helicopter support. It was not a phone call that Mansoor liked to get, but he always honored the commitment made by his boss.

Cripwell took the right people to fix the problem, and that normally consisted of one or two US contractors and a couple of engineers from the Ministry of Electricity. An energetic engineer from the embassy IRMO group by the name of Kevin Thomas was a frequent passenger. The use of the CG's helicopters highlighted the importance of fixing the problem but also physically expedited repair of the damage by as much as a week or even months in some cases. It was good to fly with Petraeus's helicopters because there were always two trailing Apaches to provide security. Downed power lines were normally in locations that required additional security.[5]

One pylon or tower that became the subject of many morning meetings over the course of several months was "Tower 57." This was a critical tower located south of Baghdad in a difficult area that seemed to be the target of insurgents whenever they wanted to bring down the electric grid. Many trips were made by helicopters to deliver material and personnel to the area. Petraeus personally got involved in the details of keeping that tower erect. When the time of the repair turned into several months, he took on the challenge and made it his personal task. Tower 57 became a nightmare for Cripwell and the electricity section of the Fusion Cell. It virtually became the EFC battle cry in the fall of 2007.[6]

The Energy Fusion Cell was physically located in Freedom Towers across the street from the engineers' compound. Freedom Towers was the tallest building in our section of the Green Zone and seemed to be a magnet for rockets and mortars. One unfortunate rocket attack killed two senior medical officers in early 2007. They worked at the military hospital located adjacent to Freedom Towers. A rocket landed next to them as they returned from the gym. It was a freakish and unlucky attack. The rocket landed close and without warning.

The damage from rocket and mortar attacks changed significantly after a radar system was installed in mid-2007. I do not remember a single casualty from incoming rounds once the radar system was installed. The system gave us about a six-second warning before the rockets or mortars landed. We were instructed to immediately seek cover or hit the ground when the alarm went off. The system proved quite effective.

Insurgents tended to launch their rocket attacks during daylight hours. Cripwell and the rest of us learned to make our phone calls to loved ones late in the evening to spare them the anxiety of knowing we were under attack. It only took one telephone call while sitting under his desk, with mortars hitting the building, to get Cripwell to start making the night phone calls part of his standard procedure.

The staff on the Energy Fusion Cell consisted of US, UK, and Iraqi military. There were a couple of Iraqi civilians: one Iraqi representative from the Ministry of Oil, and one from the Ministry of Electricity. The oil ministry representative was Ali Al Fatal, son of Hussain Al Fatal. Hussain was the Distribution Company director general who was assassinated in October 2004 after taking his daughter to the school bus stop. It was a real joy to work closely with Hussain's son and mentor him during some very difficult times. Like his father, Ali was a talented technocrat with excellent English-language skills. We continue to be friends today.

Colonel Paul Dietrich, Gary Vogler, and Captain Richard Fritzley traveling to the Bayji Refinery in April 2007. Photo courtesy of author.

About four or five Western civilians were part of the Energy Fusion Cell, assigned from the GRD contractor staff as subject-matter experts. I was one of those contractors and was assigned under the oil director, Colonel Paul Dietrich, USAR. Paul was a thirty-year employee at the Bayway Refinery in northern New Jersey, one of the oldest refineries in the United States. He was a thirty-year army reserve colonel. The Bayway Refinery had been owned by Exxon for years but had recently been sold and was currently owned by ConocoPhillips.

Dietrich had been working in the military's Civil Affairs (CA) section at the palace as their oil expert. He tried to coordinate those CA efforts through the embassy oil group but found that was a frustrating challenge. The creation of the Energy Fusion Cell offered him the opportunity to make a positive contribution in a way not afforded to him before. Paul felt so strongly about this opportunity and new direction that he volunteered for another year in Iraq rather than demobilize in early 2007.

Colonel Dietrich, Captain Fritzley, and I worked well together. Fritzley and Dietrich complemented each other quite well. Fritzley demobilized in May 2007 and was replaced by Captain Tom Brovarone, a navy nuclear submarine maintenance expert. Lieutenant Colonel Kathryn Vollert, USAR, an engineer who grew up in the California oil patch near Taft, eventually replaced Dietrich in 2008, and Colonel Jeff Freeland, USA, replaced Vollert in 2009. Jeff was a career army officer in the combat engineers with several deployments to Iraq and Afghanistan.

A key objective of the Energy Fusion Cell oil section was assisting the Bayji Refinery director general with solving his enormous operational and security problems. More details on the refinery are covered in a later chapter. The EFC oil team made several trips to Bayji during the first half of 2007 to support this effort.

It was during one of these early Bayji trips that I experienced the vulnerability that our troops were facing daily in Iraq. After landing at Camp Speicher Airbase, just to the northwest of Tikrit, we were informed that a ground movement would take us to the forward operating base, Camp Summerall, and then the remaining distance to the refinery. The number of improvised explosive devices (IEDs) along our route had increased significantly over the previous year. The movement would be at a slow, tactical pace.

We were placed in Humvees for the trip from Summerall to the refinery. These Humvees had been upgraded with armor-plated doors and flooring. However, they were still no match for the IEDs that were seriously injuring or killing our troops daily in Iraq.

This particular unit had taken additional measures and added sandbags on the floor of the Humvees to help protect them from IEDs. As we drove out of the base and toward the refinery, my knees were not far from my chin since my boots were on top of the sandbags in the back seat of the vehicle. If we did get hit with an IED, my wounds and burns would be full of sand from all the sandbags. We drove down the road close to the city of Bayji at a slow speed and made several stops until the lead vehicles could clear suspected IEDs. They turned out to be nothing more than trash along the side of the road. The trip was less than twenty miles but it took us more than an hour. These troops had no confidence in the capability of their vehicles to protect them from IEDs. They were doing the best they could with the equipment they were given.

It had been more than two years, but the question asked of Secretary Rumsfeld by army specialist Thomas Wilson of the 278th Regimental Combat Team in Kuwait in December 2004 made more sense to me. He asked, "Now why do we soldiers have to dig through local landfills for pieces of scrap metal and compromised ballistic glass to up armor our vehicles, and why don't we have those resources readily available to us?" Secretary Rumsfeld's response did not show much empathy: "you go to war with the army you have, not the army you might want or wish to have at a later time."[7]

This Humvee ride became a real eye-opener for me. There were reports that the secretary of defense, Bob Gates, was pushing for a new mine-resistant ambush protected (MRAP) tactical vehicle for quick fielding in Iraq and Afghanistan. These vehicles were being designed to withstand IEDs. Our country needed to do more for our troops than encouraging the use of field expedient measures like using sandbags.

GRD engineers and military officers were not exempt from this vulnerability. An army lieutenant colonel assigned to the procurement group had been killed by an IED in 2006 while traveling in Baghdad. The new GRD commanding general, Brigadier General Jeff Dorko, was severely injured in October 2007, just a few weeks after taking command. His convoy was hit by an IED on the north side of Baghdad and he was immediately evacuated out of the country. Dorko was lucky since his injuries were not as severe as most. He returned to Iraq after a couple of months of convalescence to complete his tour.

Fortunately, Secretary Gates had given the fielding of MRAPs top priority. Dean Popps (assistant secretary of the army for acquisitions) would advise me years later that Secretary Gates informed the Pentagon procurement team that these vehicles would be given the highest priority. A marine brigadier general set the tone with prospective suppliers early in the procurement process. He let suppliers know that Gates had authorized waivers when necessary to disregard sections of the Federal Acquisition Regulations in order to expedite the fielding of these MRAPs.[8]

Secretary Gates's leadership for this particular program would save hundreds of American lives in Iraq. The MRAPs would boost the morale of the troops and enable them to better accomplish their mission with renewed confidence in their equipment.

The multibillion-dollar procurement action was a huge, positive accomplishment by the Pentagon. An incident occurred in late 2007 that highlighted the importance of these MRAPs. I was waiting for the dining facility to open late one night for midnight rations when a Blackhawk helicopter landed next to our compound at the military hospital helipad. A few minutes later, a marine sergeant joined me in line waiting for midnight rations.

The sergeant quickly started a conversation, informing me that he had arrived on that Blackhawk from Anbar Province. He was sent to the hospital to get his ears examined after experiencing an IED attack earlier that day while riding in an MRAP.

"Tell me more about this attack. What was it like to be inside an MRAP when an IED hit?" I asked.

"Sir, if we had been in Humvees today during that IED explosion, half of my squad would be dead and the other half would be evacuated to Europe or the United States for severe burn injuries. Getting hit by an IED in an MRAP is kind of like hitting a speed bump at 40 mph. Everyone was shaken up, but no one got seriously injured. My injury was the most serious and it's because I did not have my ear protection on properly. I may have lost the hearing in my right ear," the sergeant stated.

Anecdotal stories like the marine sergeant's became commonplace as the MRAPs were fielded throughout Iraq. In August 2008, Lieutenant General Frank Helmick personally experienced such an attack in Mosul when a vehicle-borne improvised explosive device was detonated next to his MRAP. The International MaxPro Plus MRAP was severely damaged, but Helmick and the rest of the people inside walked away.[9]

There were many instances of the MRAPs saving lives, improving morale, and positively contributing to the success of the warfighter in Iraq. The troops on the ground performed much better when they had confidence in their equipment.

On behalf of all those soldiers whose lives were saved by an MRAP, thank you, Secretary Gates and your Pentagon team, and thank you to the American suppliers of the MRAP vehicles: Navistar International Military Group, Oshkosh Truck, Protected Vehicles Incorporated, Textron Marine and Land Systems, Armor Holdings, BAE Systems, Force Protection Incorporated, General Dynamics Land Systems, and General Purpose Vehicles.

Brigadier General Carew Wilks replaced Cripwell at the end of 2007. Similar to Cripwell, Carew was a UK forces engineer officer. He arrived at a time when the surge was showing some positive signs of turning things around. His focus in the oil sector was to support the export project in the south and the Bayji Refinery.

Wilks helped to identify ways to improve the Bayji Refinery power problems. Power outages on the electricity grid were the primary reason for refinery downtime in 2007. He supported efforts with the embassy electricity sector to dedicate specific generation units within the Bayji power plant to supply dedicated power to the adjacent refinery. This moved the refinery's electric power source away from the national grid and had an immediate positive impact on the refinery operations.

Another significant problem of the Bayji Refinery was its inability to vacate the huge amounts of heavy fuel oil produced as the refinery increased production levels. Wilks supported a joint effort between the Energy Fusion Cell and Joint Forces offices stateside to conduct a detailed blending study recommending optimal levels for blending heavy fuel oil with the crude oil export stream that flowed through the IT pipeline.

This blending study was welcomed by Oil Minister Shahristani and would help create the standard procedure for his refineries to follow. It would save millions of dollars each year by upgrading the value of the heavy fuel oil and improving the operations of the refineries.

Wilks departed in June 2008 and was replaced by UK Brigadier General Hamish McNinch. General McNinch spent his career in aviation maintenance, logistics, and procurement positions. He was educated as an electrical engineer and would spend much of this assignment working Ministry of Electricity issues. His Iraq assignment was his final job before his retirement in 2009.[10]

For the oil sector, McNinch continued the advocacy role for the southern export project. He recognized and educated others on the tremendous risk Iraq faced if the current dilapidated southern export system started to fail. The loss of export revenues was a huge financial risk, but there was also a potential environmental disaster that might occur from a catastrophic subsea pipeline failure.

A new initiative that McNinch supported was promoting a national energy strategy with the Prime Minister's Advisory Commission (PMAC). Thamir Ghadhban headed the PMAC organization, located about two hundred meters away from our offices in the Freedom Towers. GRD and EFC offered assistance to Ghadhban as he built his organization of experienced technocrats to more than sixty highly qualified employees within a couple years. Since Ghadhban's office was physically located close inside the Green Zone, I took the opportunity to stop by his offices quite often for social calls.

One such social call I'll always remember. Tarik Al Tahir was an Iraqi who worked with me in GRD. Tarik was part of the oil ministry team that first discovered the giant Rumaila oil field in the south in the mid-1950s. He became a director general at the oil ministry before his retirement several years prior. He mentored Thamir Ghadhban and another PMAC employee, Abdulilah Al Amir, early in their careers. Sadly, Tarik had an unfortunate accident and passed away in mid-2007.

Ghadhban recognized that one of his objectives with his new organization would be to develop an integrated national energy strategy for the government. The government of Libya had just completed their national strategic energy plan in 2006. The industry consultants who had helped the Libyans develop their strategic plans agreed to brief me on their process. The more I learned and shared with Thamir on this important initiative, the more his enthusiasm grew to develop something similar for Iraq.

We were keen to develop a more formal EFC program of support for Ghadhban's national energy strategy. McNinch was supportive but needed to take the idea higher since it had political repercussions. He discussed this initiative with the US embassy. The Iraq Transition Assistance Office (ITAO)

Three well-known Iraqi oil men—Abdulilah Al Amir, Thamir Ghadhban, and Tarik Al Tahir—in January 2007 at the PMAC offices inside the Green Zone. Photo by author.

had replaced IRMO in mid-2007. ITAO's leadership consisted of Mark Tokola of the State Department and Major General Jim Snyder. ITAO pushed back on our effort, not wanting the US government involved in anything to do with helping the Iraqis determine future strategic plans concerning oil. Nothing had changed since our prewar planning efforts; the US government was still paranoid about any perception of our influence in the oil sector.

It was disappointing that we could not help Ghadhban, but I understood the sensitivity of my government. McNinch recommended that another organization might be better suited to provide the direct assistance to the PMAC organization. The World Bank seemed to be the obvious candidate.

Simon Stolp was an Australian who had previously worked in the electricity sector of GRD. He took over the World Bank's Iraq lead position from another Australian diplomat, Andrew Goledzinowski. Andrew had worked with me during the ORHA and CPA eras. Stolp agreed that this initiative was of interest to the World Bank. We turned over our limited EFC work to the World Bank and Ghadhban's PMAC team.

The World Bank, under Stolp's leadership, brought in an international consultant group, Booz & Company, to help the PMAC organization develop an Integrated National Energy Strategy (INES) for Iraq. The INES document was approved by the Council of Ministers and issued in 2012.

I met with Stolp in Washington, DC, shortly after the successful release of the document and congratulated him and his organization for their work. He gave all of the credit to Thamir Ghadhban. Like all of the oil industry professionals that had worked with Ghadhban since 2003, Stolp also became an admirer of his talent. Ghadhban's technical understanding and his communication skills were vital for the development and eventual buy-in of the document from the Council of Ministers.

The Energy Fusion Cell overlapped the change of command from General Petraeus to General Odierno as the new commanding general for Multi-National Force–Iraq (MNF–I). This change of command took place in September 2008. The job of downsizing and eventually eliminating the Energy Fusion Cell fell upon McNinch. He departed Iraq in March 2009 and returned to the United Kingdom for his retirement.

The Energy Fusion Cell eventually morphed into the smaller Energy Services Division (ESD) under the MNF–I staff. Admiral Mark Belton was the first leader of this group. More about this will be discussed in later chapters. A key addition to the Energy Fusion Cell oil team in May 2008 was Kevin Ross. Kevin had worked with me during the CPA era as a mobilized navy reserve commander. We would anchor the oil efforts until our departure in late 2011 with the US forces.

The Energy Fusion Cell played a role in several oil initiatives. They included the installation of Pipeline Exclusion Zones (PEZ), the Bayji Refinery, the southern export project, an oil blending study, and a national energy strategy. Our most important role was providing the commander with unity of effort to support his security mission. As Petraeus often said, it was about security, security, and security.

CHAPTER 16

The Bayji Refinery and the Surge

The more I think about it, that refinery is on its last legs.
—Senior American colonel

It was December 24, 2009. Captain Christian Cook, USA, his two translators, and the mortar platoon of Headquarters Company, 2nd Battalion, 16th Infantry Regiment, were waiting at the Trebil border crossing between Jordan and Iraq in the western desert of Anbar Province. Along with Captain Cook were elements of the Marine Corps Military Police and the US Marine Corps Port of Entry Training Team (POETT) and a US marine officer who worked with POETT in Trebil.[1]

It was a cold December day in the desert with temperatures of about forty degrees. The skies were clear and there was a cold wind off of the desert. Two marine CH-46 helicopters had delivered Captain Cook's team and the marines to an area close to the border crossing. The aircraft flew back to a secure refueling station to wait for a call from Cook's team. Cook was now located in the no-man's-land area between the two passport control checkpoints, separated by a couple hundred meters.

Captain Cook was a US Army infantry company commander with the responsibility of securing the Bayji oil refinery in northern Iraq, about four hundred miles away from this border crossing. He had flown into the US air base at Al Asad in Anbar Province earlier in the day with army Blackhawk helicopters to execute Operation Righteous Return. They linked up with the marines who had security responsibility for western Iraq. The marine aircraft were used to bring everyone to the Jordan border crossing.

About two-dozen Americans were on the ground waiting for their package to arrive from Amman, Jordan. There was a lot of truck traffic going through the border but very little car traffic at this remote border crossing. It was Christmas Eve.

The cell phone system had chosen an inopportune time to fail. They had no communications with the package. Cook and the others had been waiting for more than an hour and were getting anxious. The package was an Iraqi, Dr. Ali Obaidi. Obaidi was the highly successful Iraqi director general of the Bayji oil refinery who had departed Iraq three months prior and had refused to return for fear of his life.

Without Dr. Obaidi's presence at Bayji, things were rapidly deteriorating. There had been five US Army company commanders at Bayji over the last three years working closely with the doctor. The previous four had seen huge improvements in refinery operations during their command. Captain Cook worried that the collapse would happen during his watch. He felt ownership of this Iraqi problem because he knew the impact to the security and his fellow American soldiers. He fretted that everything his predecessors had achieved would quickly disappear while he was in charge.

Just five weeks before this border operation, Obaidi sat in my northern Virginia living room with Robert Maguire and myself, discussing how he could safely return to Iraq. General Petraeus had phoned him from the CENTCOM headquarters in early November 2009, encouraging him to return. His extended absence was negatively affecting the operations at his refinery and the security mission of the US military supporting his refinery. Obaidi was trying to manage his refinery from Amman, but it was not working. His wife and children had been living in Virginia for a couple of years and he visited them over the Thanksgiving holidays. It was during this visit that Maguire suggested a plan for escorting Obaidi through the land border crossing at Trebil rather than the Baghdad Airport. The US military had also been looking at a crossing at the Syrian border, but decided that Trebil would be better.

On the day of this operation, Dr. Obaidi got a friend to drive him from Amman to the border. His plan was to stop and return to Amman if he did not see Captain Cook at the border. He was in the back seat of his friend's car trying to get a cell phone signal when the vehicle arrived at the border.

Captain Cook and his team went to meet the Iraqi border patrol colonel to ensure that Obaidi's processing was expedited upon his arrival. He kept checking his phone signal during the meeting with the Iraqi colonel. There were no communications and he was running out of time. The helicopters had a short window. The crews could not continue waiting for Dr. Obaidi much longer. Captain Cook, the marine major, and the Jordanian liaison officer left the meeting to walk to the actual crossing point. Cook left his translator, Hadi, and some of his men with the Iraqi colonel, buying his patience by continuing a previously initiated cordial discussion on the future of Iraq.

Cook started pacing back and forth in the no-man's-land. There was still no phone signal and no sign of Obaidi. As an infantry commander, he started looking at the security exposure of his men as the large trucks passed by them at the border crossing. It was another reason for concern.

Then, when it appeared that Operation Righteous Return would be a failure, his nonfunctioning cell phone started to ring. He quickly answered.

Obaidi greeted and told him that he was sitting in a line of vehicles not far away. The Jordanian liaison officer jumped in a pickup truck and sped to Obaidi's location and escorted him to the front of the line and to Captain Cook's location.

As they drove toward the Iraqi checkpoint, Dr. Obaidi recognized the US Army team with Captain Cook. Cook approached the vehicle and saw Obaidi get out with a big smile on his face.[2]

Cook greeted Dr. Obaidi when he exited his vehicle.

"Good to see you, Captain Cook. I cannot express in words my appreciation to you for meeting me like this. If I tried to return to Iraq through an airport, I would be a dead man," Obaidi replied. The two men briefly hugged each other in a show of relief. It was over. Operation Righteous Return had captured success from the jaws of failure.

Captain Cook wanted Obaidi to leave immediately, before the helicopter pilots lost their flying window. However, Dr. Obaidi requested that they wait until he could get his passport properly stamped by the Iraqi authorities, so that his entry was legal.

The marine major drove Obaidi to the Iraqi passport building and escorted him into the checkpoint. There were only a few people in the building. Obaidi was rushed to the head of the line. The Iraqi passport official checked the computer and notified a fellow Iraqi security officer of a problem.

The prime minister's office, Oil Minister Shahristani, and the oil ministry's security director had all assured Dr. Obaidi that the records had been corrected. He could return to Iraq without any fear. They told him that his name had been cleared and there would not be a problem at passport control. They were wrong.

Obaidi had shared his concern with me the previous month while in Virginia. If he had flown into the Baghdad Airport where there was no US military presence, the Shia passport officers would have detained him. The immigration annotation on his computer record would have placed him in detention. Detention in Iraq could be fatal for a well-known Sunni leader like Obaidi. Three other oil ministry Sunni leaders had been kidnapped over the previous three years. He did not want to be the fourth.

Dr. Obaidi had warned the marine major that there might be a serious issue with passport control before they entered the building. He now alerted the major that help was needed. The major quickly turned on his marine charm, raising his voice and intimidating the passport officials. The marine was a big man and could easily intimidate people. It worked.

The young Iraqi official gave into the pressure and stamped Obaidi's passport. He leaned over to Dr. Obaidi and quietly warned him that the computer indicated that there was a serious problem. He advised that Dr. Obaidi get it cleared up in Baghdad as soon as possible.

Everyone hurried to the assembly area to meet the CH-46 helicopters and get ready for departure. The pilots found a black Star Wars–looking helmet for Obaidi's head to go with his dark business suit and tie. Everyone loaded on the large helicopters and took off for the air base at Al Asad. Obaidi and his army escort linked up with the Blackhawk helicopters for the connection flight to Bayji, arriving after dark. It was a long trip that night, but a celebration had started and would continue for a couple of days.

Obaidi contacted his Iraqi security team upon arrival and they greeted him as a returning hero. The local political leaders sent their greetings for his return. The Sala Ad Din Provincial television station ran text at the bottom of the screen for a day congratulating him on his return.

The next morning was Christmas. Dr. Obaidi knew the significance of Christmas to his American friends. He had contacted his people at the refinery several days before his return to prepare for a feast. It was a Christmas feast, and it was a celebration. Dr. Obaidi had been out of the country for more than three months. The Americans and the Iraqi refinery employees were celebrating his return.

He had his people erect a Christmas tree and prepare a feast. It was a celebration that would be remembered. Following an American military tradition, the officers served American food and Iraqi-prepared food to the US soldiers and a few select refinery employees. The battalion commander, Lieutenant Colonel Ken Gonzalez, USA, and his staff participated in the celebration.

What was the significance of this event? Why did the US military go to great lengths to assist this Iraqi refinery manager in returning to his refinery? Why could he not pass through his country's passport control without being arrested? And why could a Sunni leader be so popular in his community while fearing for his life in Baghdad?

To better understand this story and the importance of Dr. Obaidi to the mission of the US military in northern Iraq, let's start from the beginning, in early 2007.

The Bayji Refinery was operating at about 25 percent of its processing capacity of 310,000 barrels per day of crude oil. This refinery provided more than 50 percent of the country's petroleum product domestic production when operating at capacity. It had been operating at the 300,000 level about fifteen months prior, in 2005. A multitude of security issues had occurred that contributed to the decrease in its performance.

The most significant security incident was the attack on the Al Askari Mosque in Samarra, less than eighty miles to the south. This attack occurred in February 2006 and was conducted by Al Qaeda elements. No one was killed, but it was an attack on a sacred Shia mosque. The event prompted revenge attacks by the Shia community on Sunnis, and the Sunnis responded accordingly.

Within the Bayji Refinery, the minority of Shia workers and engineers became the target of the Sunni employees and locals with ties to Al Qaeda. The situation became grave and out of control. The oil ministry leadership in Baghdad was forced to relocate the Shia employees from Bayji to Baghdad or Basra. Many of these Shia engineers held important positions at Bayji and were essential for keeping the refinery operating. Nevertheless, they had to be relocated for their own safety.

The increased sectarian violence and influence by Al Qaeda in the north carried over to inside the refinery fence. Al Qaeda heavily influenced the leadership and morale at the refinery. They were in control in early 2007. The refinery manager was a local Bayji native who had worked at the refinery his entire life. He was not willing to do anything to jeopardize the security of himself and his family. Al Qaeda was making decisions on who should be hired and who should be promoted to the key positions.

Captain Quentin Coleman, USA, knew better than anyone just how serious the situation was getting in early 2007. He investigated and documented many of the problems involving Al Qaeda's influence in controlling refinery operations, the movement of products, smuggling, the black market, and the distribution of the money among Al Qaeda and corrupt parties. Captain Coleman was the company commander of Bravo Company, 1st Battalion, 505th Infantry Regiment, 82nd Airborne Division.

Coleman was located on a forward operating base (FOB) less than twenty miles away from the refinery. His unit was responsible for security in the area in the vicinity of the refinery. Coleman's analyses were shared up the chain of command and eventually with Prime Minister Maliki.

The prime minister and oil minister had known that they had a serious problem at Bayji, but Coleman's reporting highlighted factually just how serious the problem was.

The oil minister had even considered shutting down the refinery at one point but decided that he needed to send new leadership and attempt to fix the problems. The right person was working inside the oil ministry headquarters in Baghdad at the time. His name was Dr. Ali Obaidi.

Obaidi was born and raised in Baghdad and graduated from Baghdad University in 1978. He started working at the Doura Refinery in Baghdad two months after graduating college with a chemical engineering degree. He went into military service in 1980 for two years during the Iran-Iraq War to teach soldiers how to safely handle chemical weapons.[3]

After the army, Obaidi went to the Bayji Refinery in 1983 during its initial construction and start-up. He remained at Bayji until 1986, when he applied for graduate school to get his PhD in chemical engineering. His first choice for grad school was the Massachusetts Institute of Technology (MIT). MIT accepted him, but Saddam's government did not allow him to attend. Instead, he was sent to Moscow for schooling, and he returned in 1992.

Obaidi would move to positions in the research department, the Bayji Refinery, and the Ministry of Oil planning department before 2003. He was sent to the Bayji Refinery in early 2004 as their production manager, the number-two person in charge at the refinery. It was during this time that the refinery undertook a massive maintenance project. They worked with KBR and the US Army Corps of Engineers. Obaidi played a key role in this work.

The refinery hit record production levels in late 2004 and 2005. Obaidi returned to Baghdad and worked for Dr. Ahmed Barifkani in 2005 and was in charge of refinery operations for the ministry. Dr. Obaidi was the most qualified person to fix the problems at Bayji. He was appointed in February 2007 as the director general of the North Refining Company (NRC), headquartered in Bayji.

The timing of this appointment could not have been better. It coincided with the timing of President Bush's appointment of General Petraeus. Fixing the problems at the Bayji Refinery was a high priority for both Dr. Ali Obaidi and General Petraeus.

Retired General David Petraeus was asked in November 2015 to reflect on the importance of the Bayji Refinery to Iraq during his tenure as commanding general of Multi-National Force–Iraq. His answer: "The success at Bayji was of enormous value, operationally and symbolically, and, of course, in terms of the importance of getting Iraq's largest refinery and energy producing complex functioning again."[4]

The Bayji complex consisted of a large refinery and an adjoining power plant, which were originally constructed in the early 1980s. Bayji is located about 150 miles north of Baghdad, not far from Camp Speicher, which was used as a large US military base close to Tikrit.

The refinery was strategic in 2007 because it was the largest refinery in Iraq, with enough capacity to supply more than half of the domestically produced gasoline and diesel products. ISIS considered it strategic seven years later.

Bayji received a lot of publicity and was considered strategic to ISIS in 2014 and 2015. The Islamic State captured the refinery and held it until Oc-

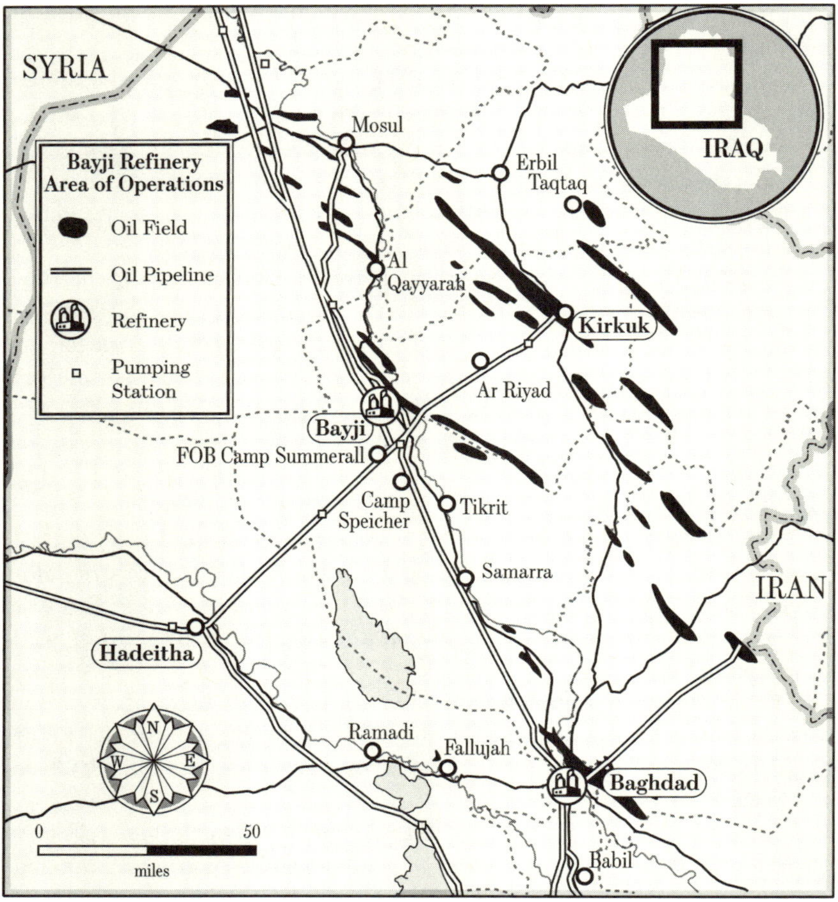

Bayji Refinery Area of Operations as of 2007.

tober 2015. The Iraqi army successfully defended it for most of 2014, but it eventually fell.

The refinery was the largest employer in Salah Ad Din Province in 2007, with a payroll of about five thousand employees. It also provided income and jobs to the local economy.

Bayji was initially constructed along the Iraq-Turkey export pipeline that runs from the oil fields in Kirkuk west to Bayji and then north to Mosul and to Turkey. There are several pipelines that carry crude oil from Kirkuk to the Bayji area. Most were old, corroded, and poorly maintained. The Iraqis only repaired the pipelines when severe leaks were detected.

The largest crude unit within Bayji and its associated upgrading units looked very familiar to me upon my first tour of the refinery. The same companies that constructed the Mobil Oil/Saudi Yanbu Refinery where I worked in the 1980s and early 1990s provided the process design and construction in Bayji. Yanbu was constructed in 1982 and Bayji was constructed in 1984.

Besides the Bayji Refinery, Dr. Obaidi also had several smaller refineries in the north in his portfolio. They included refineries at Kirkuk, Hadeitha, Sanyiah, Qayyarah, Kisick, and the underground refinery just north of Bayji. Yes, Saddam had actually directed the oil ministry to construct a refinery in the mountains, underground. He wanted to hide it from enemy aircraft. It never operated after 2003 and would be disassembled and used elsewhere in the country.

Dr. Obaidi showed up for work on February 7, 2007. He found security at the refinery to be poor. He was sitting at his desk on the second day when his administrative assistant entered his office with an anxious look on his face. "Dr. Obaidi, please sir, there is a visitor for you. I beg you. Just listen to him. Please do not irritate him."

Obaidi received this intimidating visitor. It was the senior Islamic State representative, a local affiliate of Al Qaeda. His directions to Obaidi were simple: "You will hire no Shia at this refinery. You will only hire Sunni here."[5]

The Islamic State representative advised Dr. Obaidi that his people had checked him out and that they found Obaidi acceptable. They could work with a Sunni manager with a lot of credibility at the oil ministry headquarters in Baghdad.

The Al Qaeda thug carried himself as any goon would. He had little idea of the technical requirements of operating a refinery. He was only interested in ruling the refinery through intimidation. It was a very dangerous situation at the time. Al Qaeda was ruthless and powerful. Obaidi knew that he had to latch on to a bigger and stronger group to support him against Al Qaeda. The local Obaidi Tribe and other tribes in the area were not as powerful as Al Qaeda. He needed a powerful tribe that would support him in cleaning up the mess he was given. That tribe would be the US Army.

The Americans visited the refinery during Dr. Obaidi's second week in the job. He quickly shared his concerns with the army officers who visited him. They included Captain Quinton Coleman and Colonel Harris of the 82nd Airborne Division. "The initial visit was good in that the Americans seemed to be sincere in their willingness to help me fix both the operational and security problems I was faced with," Obaidi shared years later.[6]

The following quotation from Obaidi is indicative of the intimidation that Al Qaeda created:

> During this initial visit by the US army, I kept pressing the call button for my assistants to bring in refreshments for my guests, but my assistants refused to enter my office. I became irritated with them, and after the Americans departed, I called them into my office and chastised them for not providing refreshments for my guests. Their response let me know the level of intimidation that Al Qaeda had imposed on all workers at the refinery, including my closest administration support. They were afraid that word would get to the Al Qaeda leadership that they helped the Americans. I let them know that everything that occurred in our section of the building had to be kept extremely confidential. We had to look out for each other. I would look out for their safety and they would look out for my safety and support me. I asked if anyone was uncomfortable with the new arrangement. I was willing to reassign them to another section in the refinery if they could not support these rules. There was never a problem again with my staff.

Captain Coleman and his superiors continued to visit Obaidi every couple of days. He developed a plan for assisting the refinery with security issues. Meanwhile, GRD and the Energy Fusion Cell were tasked with helping Obaidi solve his technical problems. One of GRD's contractors with more than twenty-five years of project work in the oil industry volunteered to move from Basra to Bayji. His name was Robert Carruthers.

Robert was a UK passport holder from Scotland who had been doing project work with Foster Wheeler throughout the world and recently in Iraq. He had been working in Basra since 2005 under the GRD oil program. This was his first experience working with frontline US Army combat troops. It was a real eye-opener.[7]

Carruthers's first trip from Camp Summerall to the Bayji Refinery was in a Humvee and was his first ride in a military Humvee. Captain Coleman called his unit together for a serious review of their detailed plan for the trip. He reviewed probable attack areas, procedures to follow in case of attack, and communications protocol during the trip.

Coleman then called everyone together for a solemn prayer. This caught Robert's attention. He wondered why he had volunteered for such a dangerous mission. He was not a particularly religious person and had never experienced such religious activity in his professional life before this. He now understood the American military maxim that there are no atheists in a combat unit during war.

The trip lasted about sixty minutes, traveling less than twenty miles. Carruthers got his first glimpse of a combat unit in action from the back seat of a Humvee. Fortunately, nothing happened during his first or subsequent trips to the refinery. He became a trusted advisor and friend to the soldiers of Captain Coleman's unit and the four commanders who followed over the next few years.

Carruthers also quickly became a trusted advisor to Dr. Obaidi. He was primarily used to troubleshoot operations and maintenance issues. I frequently spoke to Robert from Baghdad to provide him with reach-back support, as needed.

Carruthers remained as one of our contractors until 2009, supervised by the Energy Fusion Cell. When the EFC downsized in 2009, he remained at Bayji, working for a logistics contractor, but he continued to help Obaidi with refinery technical problems. He eventually took a job in Basra in 2010 for an international oil company.

The first operational problem that Obaidi solved was keeping the flow of crude oil constant from the oil fields in Kirkuk. Attacks on the pipelines between Bayji and Kirkuk limited the amount of oil feeding the refinery. He called in the tribal leaders of the more powerful tribes in the area. The first was from his own clan, the leader of the Obaidi Tribe. He explained to them the importance of keeping the oil flowing so that he could operate the refinery. He learned that their motivation for attacking the pipeline was that they thought all of the oil was going to Baghdad, and nothing for the north.

Dr. Obaidi assured the tribal leaders that if they would stop attacking the pipelines and, in fact, protect the pipelines, then he would hire the younger members of their tribe at the refinery. The agreement worked. The attacks on their crude oil pipeline stopped.

One visitor he received in early 2007 was a former general officer in the intelligence branch of the Iraqi army. He had been a senior member of the Baath party. The former general literally cried in front of Obaidi, begging for a job. He had lost the respect of his family and friends since 2003 and he desperately needed a respectable job. Obaidi knew that he could not hire this former high-level Baathist general officer. However, he agreed to hire his sons at the refinery. Such empathy, diplomacy, and wisdom went a long way in keeping his refinery operational and maintaining stability in the community.

Colonel Harris visited Obaidi one day in late May 2007 to discuss security issues. At the end of his discussion, he mentioned that he had something that was extremely confidential and could not be shared with anyone. There would be a VIP visitor from Baghdad in early June. The visitor would be the commanding general, General Petraeus.

The visit turned out quite well. Colonel John "J. R." Martin extended an invitation to me to join Petraeus's group. J. R. had been the executive officer for Jay Garner during the ORHA era of 2003. He was a West Point classmate and colleague of General Petraeus who kept coming out of retirement to assist his classmate when called. It was a one-day trip on June 9, 2007. We flew from the Green Zone helipad early in the morning and landed at the helipad next to the refinery. We met in Dr. Obaidi's office, took a short tour of the refinery and distribution areas, and had a great lunch. After lunch, we visited the new US Army FOB being constructed inside the refinery fence.

With Dr. Obaidi's permission, the US Army moved a company-sized unit to a base inside of the refinery. This gave the military 24/7 presence on the refinery site and enhanced their ability to better support Obaidi. Petraeus visited the new base and spoke to the troops, emphasizing the importance of their mission to support the refinery.

General Petraeus then conducted a "push-up" contest with the younger troops. I observed the last young soldier quitting after about ninety push-ups, and Petraeus did a few more just to satisfy all observers that he was the winner.

One of the accomplishments of our visit was agreeing to a plan for getting Obaidi's wife and two daughters out of Jordan and to the United States so that he could continue his work without fearing for their lives. Dr. Obaidi had moved them to Jordan a few months earlier for their safety, but their visas had expired. They needed a new home. General Petraeus asked Obaidi if he wanted his family to go to the United States.

Dr. Obaidi looked the general in the eye and said, "General Petraeus, this is my dream."

The general directed that a letter be drafted for his signature requesting that the US ambassador in Jordan assist Dr. Obaidi's family. A young FSO assigned to the local Provincial Reconstruction Team took the lead to expedite that initiative. His name was Kit Redmer. Obaidi's family was successfully relocated to Fairfax, Virginia, within a couple of weeks.

The local Lutheran community in Fairfax, led by a woman named Karen, ensured that Obaidi's family received everything they needed. Karen became the ambassador of goodwill for the Obaidi family. She enrolled the children in school, helped find a job for Obaidi's wife, and drove them wherever they needed to go for several months. When Dr. Obaidi visited his family in Fairfax after a year, he sought out Karen to thank her for all of her assistance. He will never forget Karen's words to him after he thanked her: "I am the lucky one, because God sent me your wife and children to help."[8]

The summer of 2007 passed by quickly and Captain Coleman rotated out with his unit. The incoming replacement was Captain Joe DaSilva, company commander of headquarters company, 1st Battalion, 327th Infantry Regiment, 101st Airborne Division. DaSilva's battalion commander was Lieutenant Colonel Pete Wilhelm, and his brigade commander was Colonel Scott McBride. The handover of responsibility was known as a relief-in-place (RIP). The two military units normally spent anywhere from a week to a month together before the old unit departed.

The handover between Captain Coleman and Captain DaSilva was crammed into a little over a week. Quentin's extensive understanding of the details of the Bayji operation was intimidating to Joe. How could he ever get himself and his unit up to speed quickly? Joe knew that he had to pick Quentin's brain as best he could in a short period and reach out to others once his predecessor departed. He reached out to a friend working at a major oil company who immediately sent him a booklet titled *Oil for Dummies*. Joe also quickly recognized that Robert Carruthers would help him fill that void of technical knowledge that he needed filled.

Joe learned from his new Iraqi partner, Dr. Ali Obaidi. The relationship between Dr. Obaidi and Joe DaSilva over the following fifteen months grew into a close friendship. Their fight to rid the refinery of the Al Qaeda influence made them trust each other, just as two soldiers sharing the same foxhole during war.[9]

Early assessments by Colonel McBride and Major General Mark Hertling were not encouraging. Shortly after arriving in Iraq, McBride told DaSilva, "The more I think about it, that refinery is on its last legs."[10]

Major General Mark Hertling was the commander of Multi-National Division North. He visited Joe DaSilva during his third week at Bayji. He called Joe aside and expressed his doubts about needing an infantry company located at an oil refinery when it was needed more elsewhere. Hertling mentioned to Joe that he did not understand why troops were needed at the refinery and asked Joe for his thoughts on the issue.

"Sir, my predecessor was convinced that our mission here was extremely important. I have only been here a very short time so I don't want to give you the impression that I am an expert, yet. Please let me give you a tour and we can talk more after we are finished," Joe told the general.

Joe showed General Hertling around the refinery, the distribution center, the oil lubes plant, the fertilizer plant, the detergent plant, and the power plant. The general saw the wheels of industry in motion and the thousands of Iraqis at work. They did not need to have that conversation about why the Bayji mission was so important. Hertling saw the answer with his own eyes.

General Hertling supported DaSilva with additional intelligence resources from the division headquarters. An army major and a civilian were reassigned to Bayji. Captain Steve Wright was sent from the brigade headquarters to also augment the intelligence capability.

A wave of attacks hit the refinery and surrounding areas in late 2007. Four improvised explosive devices (IEDs) and three vehicle-borne improvised explosive devices (VBIEDs) were detonated. One went off in Dr. Obaidi's housing area. A strategy of intimidation of the Iraqi employees by the insurgents was succeeding. There was fear in the eyes of Obaidi and his employees.

Captain DaSilva understood the magnitude of the job in front of him. He knew that he would need to reorganize his infantry company to better undertake the task. DaSilva put together an intelligence cell, an operations group, and other specialized functions within his company. He knew that a reengineered organization was necessary to address the problems. His team studied the intelligence data and developed a plan of attack.

DaSilva's company developed a response to the recent attacks, Operation Boston, named after Joe's hometown. There was a housing village adjacent to the refinery called the Czech Village. These houses were constructed in the early 1980s, when the refinery was built. Two of the main process lines were designed and constructed by a Czech company. The Czech Village was where the Czech contractors had lived during construction.[11]

The Bayji Refinery used the Czech Village to house any contractors working at the refinery. The village had been vacant for the previous year because of the security situation; no contractors agreed to work at the refinery. Al Qaeda had moved in and used the vacant village for their operations. DaSilva organized a raid on the housing compound.

Iraqi staff Colonel Adel and his Oil Protection Force (OPF) swept the village in an early-morning raid with about twenty Iraqis. Several soldiers backed him from DaSilva's unit. The element of surprise was achieved; all of the terrorists were rousted from their sleep. The village was secured with concertina wire on the perimeter and Iraqi guards to control access after the raid. This was the first of many major combined operations between the Iraqi OPF and the US Army at the Bayji Refinery.

The raid discovered a cachet of ammunition, IEDs, a suicide vest with explosives, and weapons. More than a dozen Iraqis were arrested. The operation was a success. It cleaned out what was being used as a supply depot by Al Qaeda.

Several other arrests occurred within a few weeks of the raid. Captain DaSilva asked Dr. Obaidi to identify all of the bad guys that were operating inside his refinery. Obaidi identified thirty-seven employees that he knew were part of the Al Qaeda element operating inside his organization. DaSilva's intelligence cell had been tracking thirty-four of the thirty-seven identified by Obaidi as known Al Qaeda players. All were arrested and sent to a detention facility.

This operation of removing known bad guys proved successful. Morale and discipline improved. Employees started coming to work on time. Dr. Obaidi and Colonel Adel identified the culprits, and the US Army detained them. So many bad guys were apprehended during late 2007 and 2008 that the US military nicknamed Obaidi and Colonel Adel as "Batman and Robin."

The soldiers would learn that Colonel Adel was actually a former general officer in the Iraqi air force. He piloted several missions during the Iran-Iraq War over the northern Arabian Gulf. He always made sure that he communicated with the US Navy 5th Fleet as he went on those missions. They never replied to his radio communications, but he always felt comfortable letting them know that he was a "friendly" plane flying over the Arabian Gulf. Then one day an Iranian fighter unknowingly got behind him for an attack. Radio silence was broken and an American voice warned Colonel Adel that he had enemy on his tail. He maneuvered and escaped to fight another day.[12]

The US military moved to the next threat to the refinery, corruption. A key arrest was made in March 2008. The person was the Bayji Refinery inspector general, the IG. He reported directly to the inspector general in Baghdad. The local IG had been involved in corruption activities that became apparent to both Obaidi and the US military. The guy responsible for fighting corruption was the biggest crook. He was arrested and sentenced to one year by the local courts. The *New York Times* ran a story of this specific case. Many co-conspirators were later identified and fired, an action that would have been impossible a few months earlier.

The morale of the Iraqi employees was high. I personally experienced this with one of their more competent technical employees. I had first met Dr. Riyadh in 2003 on an early visit to the refinery. He was forty years old and worked on the North refinery process. Upon my return in early 2007, he pulled me aside during a tour and begged me to help him get out of this refinery. His life was being threatened and the work environment was dangerous. A year later, in early 2008, I saw him again while taking a tour of the North refinery. I asked him if he still needed to leave. He quickly replied, "No, Mr. Gary. Life is so much better under Dr. Obaidi. We enjoy working here again."

With security and morale improving, Obaidi focused more of his efforts on increasing production. Everyone focused on helping Bayji get its production up. Joe DaSilva would get phone calls from the chief of staff to General Petraeus asking specific questions about production outages. On one occasion, he phoned Joe to inquire about a power outage that had just occurred and where the power was still out. There was a sense of urgency by everyone to help Obaidi.

Oil Minister Shahristani was impressed with the success achieved at the Bayji Refinery. Production volumes increased until the refinery operated at 80 percent of capacity at the end of 2007. The minister wanted to give Obaidi more responsibility in the north, placing him in charge of the highly corrupt product distribution system.

Dr. Obaidi resisted this new assignment because he knew the dangers of such an operation. Distribution was responsible for taking product from the refinery and moving it via trucks or pipelines to the ultimate customer. This operation reported into the Product Distribution Company in Baghdad. However, it was well known that the Bayji distribution operation was more corrupt than others.

The oil minister continued to press Obaidi to take on this additional responsibility, but Obaidi was well aware of the inherent dangers. The assassination of Hussain Al Fatal and the attempted assassination of Sameer Michael were on his mind. Both tried to stop corruption in distribution and both had failed. Eventually, Obaidi gave in to the oil minister and accepted the additional responsibility. Several years later, he felt that he had made a huge mistake when he agreed to this new assignment. The threats to his personal security increased.

Obaidi took control of the distribution operation in September 2008. He immediately set up new management systems for reporting. He assigned Colonel Adel as part of the Distribution Company's board of directors, fired employees for stealing, rewarded employees for turning in thieves, paid employees more money, introduced technology for the weighing system, and introduced perks for the good employees. The new system worked. With Joe DaSilva and the US military watching his back, he quickly fixed the local distribution problems.

DaSilva and Obaidi implemented a product tracking system that identified the fuel trucks being sent from the Bayji distribution area out to the other military areas of operation in Anbar, Ninewa, Kirkuk, Dyala, Erbil, and Sulaymaniyah Provinces. This enabled the US commanders and Iraqi governors to see how much fuel designated for their area was not arriving. Previously, there was no tracking. So, trucks would not deliver their cargoes to the correct location. They delivered to the highest bidder or whoever threatened them. The highest bidder was normally a smuggler who would illegally sell the product outside of the country.

One smuggling scheme that was identified during this period was the movement of diesel trucks from Bayji to Kurdistan and across the border into

Iran. These same trucks would then cross the border back into Iraq further to the south under an import contract with the Ministry of Oil. The ministry paid someone in Iran for diesel product that was produced at Bayji. It was a highly lucrative scheme until Obaidi and DaSilva stopped it.[13]

A product-stealing scheme was also discovered. The fuel volumes sent from Bayji to Baghdad were sent via tanker trucks that delivered to the Doura depot. The depot was located adjacent to the Doura Refinery and kept reporting volumes 30 percent below the Bayji shipping volumes. Obaidi had gained so much confidence in his new measuring equipment that he immediately focused on the trucking company. The trucks were owned by a private sector trucking company from Anbar Province. The Doura depot started inspecting the tanker trucks and found false walls and hidden valves on the trailers. This enabled one-third of the product to remain on the truck after unloading.

The US marines in Anbar used DaSilva's reports to initiate an escort mission. The marines sent a unit to Bayji to escort the fuel trucks destined for Anbar. This facilitated the delivery of fuel to the intended locations.

Military commanders and governors throughout the north became proactive to ensure that their assigned trucks delivered the correct volumes to their designated locations. The governors of all the northern provinces sent letters of appreciation to Dr. Obaidi for the increased fuel supplies to their provinces. Several paid him visits to thank him in person. The governor of Anbar visited him on three occasions to show his appreciation. The governor in Mosul highlighted that the storage tanks at his depots contained extra inventory for the first time since 2003.

Not everyone shared this enthusiasm. The criminal element started losing its substantial revenue stream. They were not prepared to accept less profit for their efforts. The criminals would target Batman and Robin at an opportune time.

Captain DaSilva and his headquarters company departed Iraq in November 2008 and returned home to Fort Campbell, Kentucky.

Seven years later, in October 2015, Major Joe DaSilva shared with me that this very important mission with Dr. Obaidi convinced him to make a career of the army. He was a West Point graduate and president of the class of 2002. His previous tour in Iraq had been difficult. During that tour, he lost several soldiers in his unit from IEDs in and around Baghdad. He intended to resign from the army as soon as he could after returning from that tour. Then, he changed his mind. His successes and job satisfaction working at Bayji had convinced him that the army was best for him.[14]

Headquarters Company, 2nd Battalion, 27th Infantry Regiment, 3rd Brigade, 25th Infantry Division replaced Captain DaSilva's company. Captain

Marcus Wright and then Captain Rich Sugg commanded this company between December 2008 and August 2009. These commanders continued supporting Dr. Obaidi to help him clean up the corruption in the northern distribution company and secure the refinery.

Refinery operations continued to improve during the winter months and an important milestone was achieved in February 2009, the start-up of the hydrocracker. Bayji was the only refinery in Iraq with a cracking unit, a complex process unit that is used in all Western refineries. It was installed in 1984 but seldom operated, primarily because of the unreliable power supply. The unit had not operated since before 2003. To prepare for the start-up, Obaidi sent operators and maintenance engineers to Jordan and Egypt for training during 2008.

The hydrocracker processed about thirty thousand barrels per day of heavy fuel oil into diesel fuel. This enabled the refinery to produce large volumes of diesel fuel, more than enough to meet demands throughout the north. The product depots were able to build inventories of diesel fuel for the first time in years.

It was during the first half of 2009 that both Dr. Obaidi and Colonel Adel (Batman and Robin) received serious backlash from their work to eliminate corruption from the oil distribution area. Attempts to successfully issue warrants against both gentlemen by the corrupt elements within the government had been thwarted by the US Army. Unfortunately, the cleverness of the criminals only increased with each attempt.

Two local Iraqi citizens were bribed to file false claims against Dr. Obaidi and Colonel Adel. They would later apologize to Obaidi, but the damage was done. Corrupt citizens could easily manipulate the nascent Iraqi legal system with no cost to them. In this case, these false claims enabled the issuance of warrants for arrest. No sheriff or local police were sent to arrest them. Warrants were normally only served when the citizen attempted to pass through a government checkpoint such as passport control at border checkpoints. The individuals were then arrested in accordance with the warrant.

Dr. Obaidi would not become aware of this particular warrant until he attempted to fly out of Erbil Airport in the early fall of 2009. He had a meeting in Amman, Jordan. His drivers took him to the Erbil Airport for the flight. As he was checking through immigration, the passport official checked the computer screen and found the warrant information. Obaidi recognized that there was a problem, but was able to convince the passport official that he was not the same person listed in the computer. "I am the director general of one of the largest companies in the Ministry of Oil; your computer must be wrong," Obaidi told the immigration official.

As Obaidi boarded the plane for Amman, he knew the truth. He could not return to Iraq without being arrested at an airport. As mentioned earlier in the chapter, the detention of senior Sunni leaders in the Ministry of Oil could be fatal. He decided to set up his offices in Amman and manage the refinery from there. Running a large oil refinery from five hundred miles away did not work. Many of his hiring and compensation programs were losing their effectiveness without his physical presence. The US military quickly recognized a decrease in the morale of the Iraqi employees and a threat to the stability with the local community leadership. They needed him to return.

Captain Christian Cook replaced Captain Rich Sugg in August 2009. Christian had worked with Obaidi for only about a month when the Erbil Airport incident occurred. He recognized the value of Obaidi to his mission shortly after taking command. He witnessed the operations of the refinery deteriorating quickly during the doctor's absence. Successfully planning and executing Operation Righteous Return at the Trebil border was the first step. Now that Obaidi was back, Cook's mission was to help him clear his name of the charges and the warrant.

On December 30, Captain Cook led a team of four large MRAP trucks, Dr. Obaidi, and his Iraqi security team to Tikrit for a meeting with the Iraqi judge. The front vehicle in his convoy was an MRAP vehicle with a massive plow that was designed to clear IEDs. They were supposed to get an escort from the Tikrit Iraqi police, but the police were on another mission. Cook made contact with the Tikrit police captain and was advised to move to the courthouse on his own. It was morning rush hour. His trucks moved slowly through the center of the city and on to the court.

Captain Cook parked the four large trucks across the street from the courthouse and escorted Dr. Obaidi into the building and to a point just outside of the courtroom.

The US Army brigade judge advocate general (JAG) met them at the courthouse. He was a US Army JAG officer who had been working on Dr. Obaidi's case since before his return to Bayji. The JAG officer recommended that he escort Dr. Obaidi into the courtroom without any other US military personnel. The optics of several armed soldiers in a courtroom would not be good. There were several members of the Iraqi press present at the time.

The judge ruled in favor of Dr. Obaidi. The Iraqis who filed the complaint were present and apologized to him for what they had done. The judge and Obaidi moved to the front steps of the courthouse to talk with the press. It was a show of confidence and faith in the Iraqi judicial system. Celebrations were short-lived. Another warrant surfaced a week later.

The next warrant was an accusation that Dr. Obaidi was financially supporting the insurgency. The origins of this new warrant came from the same people in Baghdad and the Bayji area who had filed previous complaints against him. It became obvious to many that Batman and Robin were seriously affecting the revenues from corruption of people at high levels in the oil ministry and the local Bayji mafia.

Captain Cook's unit escorted Obaidi to Baghdad in early 2010. A meeting was arranged at the Ministry of Oil with its senior security manager. This security manager had previously assured the oil minister that Obaidi's warrant had been cleared, but that was not the case. He also was known to have strong ties to Iran and a financial interest in a corrupt Anbar trucking company that Obaidi had barred from loading Bayji fuel. The security manager was part of the problem. The meeting ended quickly without any positive results.

Obaidi visited the US embassy for a meeting with Brigadier General Bob Ranck, the legal team from General Odierno's staff, and myself. The legal team was brought in to discuss options for clearing Dr. Obaidi's name. It was recognized that he would probably not get a fair hearing in Baghdad. His personal security was always at risk in Baghdad without the presence of the US military. Odierno's legal staff initiated a request to move the case to a judge in Tikrit.

Obaidi was given a court date in the spring of 2010 in Tikrit. He arrived at the courthouse for the hearing only to learn that the judge did not have the documents from Baghdad. They had been lost or stolen. Odierno's legal team recognized this maneuver and immediately remedied the problem. They had maintained copies of all the pertinent documents. It did not surprise them when they heard of the lost documents. They made a copy of their file and had it hand-delivered to the Tikrit judge by a US Army JAG officer. A new trial date was established and Obaidi's name was cleared.

Obaidi reinitiated his hiring program immediately after returning to Bayji at the end of 2009. His policy of hiring one person from each family paid significant dividends in security and stability in the area. "Give them a job and they will stop attacking the refinery," Obaidi said many times.

Dr. Obaidi and Colonel Adel had decided to start bunking down many nights at the same camp as the US military inside the refinery. The risk to both men had increased and they felt most secure with the US Army. This facilitated a great working relationship between the senior Iraqi leadership and the US Army at the refinery. Evening social highlights included a game of dominoes with Obaidi, Colonel Adel, Robert Carruthers, and Captain Cook. Dominoes, cigars, soccer on the TV, and Iraqi sweet tea contributed to an atmosphere of camaraderie among these men. Dominoes and talking business were how the days ended in 2009 and 2010.

Robert Carruthers and Dr. Ali Obaidi playing dominoes with US forces during the evening at the Bayji Refinery. Photo courtesy of Carruthers.

Dr. Obaidi knew that his future in Iraq was dependent on the support he received from the US military. He started advising me and others that he had to leave Iraq before the US military departed. As the months passed in 2010, it became clear to many of us in Baghdad that the Obama Administration would not succeed in keeping a sizable force in Iraq. Certainly, the size of the force would not be enough to maintain a company-size unit at the Bayji Refinery.

Captain William Sitze replaced Captain Christian Cook in May 2010. Sitze commanded Alpha Company, 2nd Battalion, 16th Infantry Regiment. Captain Sitze and his unit stayed for a short time before being replaced by Captain David Webb of the 1st Battalion, 27th Infantry Regiment. There was a quick handoff and Webb stayed until September 2010.

The turnover of US military commanders and units at the Bayji Refinery created concern about the future. Dr. Obaidi knew that he needed to join his family in the United States. He applied for early retirement from the Ministry of Oil and it was granted by the prime minister during the second half of 2010. Prime Minister Maliki did not want to lose such a valuable leader at the Bayji Refinery, but he recognized that Obaidi would probably leave once the American forces departed Bayji.

On August 16, 2010, Dr. Obaidi retired and departed Bayji for Amman, Jordan. He immediately applied to the United Nations for refugee status so that he could join his family in northern Virginia. He joined his family on November 1, 2010.

Mr. Abdul Gafour replaced Obaidi as the director general of North Refining Company. Gafour had been educated in the United Kingdom and had strong technical credentials, but he lacked the leadership skills of his predecessor. With the departure of Obaidi, followed by the departure of the US forces from Bayji, it was only a matter of time before there would be an impact to Bayji operations.

A major attack on the refinery took place in the spring of 2011. I was dispatched from Baghdad to Bayji to assess the damage and see if there was anything the US forces could do to assist. Dathar Kashab joined me on the trip. Dathar had retired from the Doura Refinery at the end of December 2010 and welcomed the opportunity to accompany me on the trip to Bayji.

We met Mr. Gafour at the refinery and he gave us a tour of the damage. The damage was severe, but it could have been much worse. The attackers had obviously underestimated the explosive charges required to do the damage they intended. Their charges did not explode on some units while on others the pipe deflected the explosive force so that the damage was minimal.

The saboteurs had entered the refinery at an opening along a remote area of the fence. They had good information about the layout of the refinery and how best to enter and depart. One employee on the early morning shift was killed. We assumed that he may have detected them in the dark and tried to approach them. They placed their explosive charges on important units of the North refinery and departed.

The engineers at the refinery shared their plan to repair the damages. It was a good plan. Both Dathar and I agreed with their strategy. It would return several units to production as quickly as possible. That part was easy and manageable. There was something that would be much more difficult to fix. The look of fear in the eyes of the competent engineers and leaders at the refinery had returned. It was the same fear that I had sensed in early 2007 from Dr. Riyadh and others during that tour.

Dathar also detected the fear and lack of confidence by these managers. He had known them for much longer than I had. Director Gafour was a good man, but he was not able to fill the shoes of Dr. Obaidi. The local Oil Protection Force was not able to fill the security vacuum upon the departure of the US infantry company.

Bayji continued to operate at about 50 percent capacity over the following three years. It did not achieve the success of 2007 through 2010, but it also did not fall back into the dismal operations of 2006. That is, until ISIS moved south from Mosul in June 2014. Many of the engineers and managers escaped the refinery for Baghdad or Erbil, awaiting an opportunity to return to their refinery.

Dr. Obaidi stayed with his family in northern Virginia. He refused to return to Iraq for more than five years. He visited Amman, Jordan, on several occasions and maintained communications with his friends in the Ministry of Oil. His wife and two daughters had become US citizens, and he also achieved that status in February 2016.

It pleases me to have helped Dr. Obaidi and his family escape the threats of Iraq and become US citizens. I consider him a close friend. We share lunch and stories periodically in northern Virginia. However, I am also reminded of my conversations with Thamir Ghadhban back in 2003 and 2004. Ghadhban always lamented how the wars of Iraq over the last thirty years had forced a "brain drain" of the best and brightest Iraqis. At great cost to the future of their country, many clever and smart Iraqis escaped Iraq. Many went to Europe, North America, and other locations in their desire to live a secure and prosperous life, away from the threats and wars of Iraq.

The microcosm of the Bayji industrial complex and the northern region highlights something else. The only way for this to have been a sustainable success story would have been for a much longer commitment by the United States. The short-lived partnership at Bayji between the US military and Dr. Obaidi was a success, but it could not be sustained without a longer US commitment.

CHAPTER 17

The Southern Export Project

This export project is not only the most important oil project, but the most important infrastructure project completed in decades.
—Oil Minister Abdul Karim Al Luaibi

In 2007, Iraq's southern export infrastructure consisted of Al Basra oil terminal (ABOT) and the smaller and damaged Khor Al Amaya oil terminal (KAAOT). Both were old and in poor condition. Their combined capacity for loading tankers was about 1.7 million barrels per day and almost all was at ABOT. Iraq's exports and associated revenue depended on ABOT. There was no backup and no other export channel in the south.

ABOT was constructed in 1975 and was located about thirty miles south of the southern Iraqi beaches. It was located in the deepest navigation channel of the northern Arabian Gulf, the only channel deep enough to handle very large crude carriers (VLCC). KAAOT was constructed in 1960 and was located further to the east of ABOT in a shallow channel. It could not handle VLCCs.

Any successful attack, industrial accident, or equipment failure on ABOT translated into an immediate negative impact on Iraq's fragile financial posture. The Iraqi oil protection force and US Navy repulsed an attack in early 2004 from two speedboats full of explosives piloted by suicide drivers. An industrial accident in 2006 destroyed much of the KAAOT facility.

Iraq had attempted three different single-point mooring (SPM) projects in the south since the early 1990s to add redundancy to its southern export infrastructure. The first project failed after the contractor filed for bankruptcy in the 1990s. A partially completed SPM was still stored somewhere in Southeast Asia as part of that initial project.

The second project was started in 2005 in partnership with the government of Japan and their Japan International Cooperation Agency (JICA). JICA provided low-interest loans and technical support to construct a subsea pipeline with one SPM. Contract issues between the oil ministry and the JICA contractors delayed that project. As of the end of 2016, the project had not been completed.

The third SPM project is the subject of this chapter. It was initiated in 2007 and started up less than five years later, in the first quarter of 2012.

Single-point moorings (SPMs) in fabrication yard during August 2011. The light gray equipment on the left is a pipeline end manifold (PLEM) that is installed on the seabed under the floating SPM. Photo by author.

Iraq exported 3.3 million barrels per day out of the south during the last two months of 2015, twice as much as at the end of 2011. The Kurds also exported about 0.6 million barrels per day out of northern Iraq during the same two-month period. The new SPM project enabled Iraq to double exports out of the south, provide billions of dollars to the Iraqi treasury, and propel them into the number-two oil production country within OPEC.

The oil minister spoke in early 2012 during the SPM project start-up ceremony: "This export project was not only the most important oil project, but the most important infrastructure project completed in decades."[1] This chapter will discuss the history, challenges, and successes.

In early 2007, General Petraeus signed a letter to Prime Minister Maliki identifying the southern export infrastructure as a serious threat to their national security. This fragile export infrastructure provided 95 percent of the revenues to the government. He offered to assist the Iraqis in installing redundant export infrastructure, specifically SPMs. The execution was assigned to the Corps of Engineers at GRD in Baghdad. US Navy Captain Richard Fritzley and Lieutenant Commander Tim Spicer were the US government leads. As a contractor, I provided technical assistance and advice.

Progressing with the initial feasibility study within the US channels was not without obstacles. Although the project had significant top-level military coverage by General Petraeus, the embassy was not totally onboard.

The first challenge was an easy fix. The embassy oil team challenged the wisdom of installing SPMs and not fixed export facilities. They claimed that no country in the Arabian Gulf used SPMs; they only used fixed offshore facilities. The facts refuted this position. An industry ship captain advised that there were more than four dozen SPMs operated by Arabian Gulf countries at the time and only Iraq and Bahrain were not using SPMs. Bahrain was installing one.

The next challenge by the embassy oil team was a bit more troublesome. I received a phone call from Dr. Les Dixon early one morning in May to come to his office. Les was the number-two person in GRD, a senior executive service employee with the Corps of Engineers. Upon entering his office, he said to me that the export project was dead because the embassy did not want it. He proceeded to show me an email from the embassy oil group saying that the Ministry of Oil had no interest in pursuing an SPM project with the US government. I immediately recognized the inconsistency of the email with comments that were previously made to me by Deputy Minister Ahmed Al Shamma.

I looked at Les Dixon and said, "This is obviously wrong or at least inconsistent with the discussions I've had with the oil ministry's leadership."

Dixon responded that the US embassy was our customer and if our customer did not want the project, then we would not do it. Les was an excellent manager and we worked well together, but this was one instance of disagreement.

It was time to bring out the "defibrillator" to keep life in this project. It would not be the last time that this project would need resuscitation over the next few years.

"Who is going to advise General Petraeus that his SPM project has been canceled?" I asked.

After more discussion, it was agreed that I would be allowed to take US Navy Captain Tom Brovarone to get Deputy Oil Minister Shamma's comments about the embassy's recent email to kill the SPM project. Captain Brovarone had recently replaced Captain Fritzley as the oil lead in GRD. Deputy Minister Shamma not only refuted the comments in the email but also emphasized the need for the project to continue expeditiously.

It took several months for the US embassy to get onboard and support the SPM initiative. I sensed that the resistance was primarily because the Japanese had an SPM project in play and our guys did not want to compete with the Japanese. However, there was a huge difference between the objectives of the Japanese and those of General Petraeus. The Japanese saw their project as an oil project for their companies to participate in, while General Petraeus focused on security. He considered it an Iraqi national security project, not an oil project.

As a first step, GRD agreed to perform a feasibility study. The security situation was so poor in 2006 and 2007 that all three of the oil sector engineering firms under contract with GRD wanted to complete their existing projects and leave Iraq. They were not looking for any new work. Foster Wheeler agreed to do the study as long as they did not have to send any of their employees inside of Iraq to complete the work.

The feasibility study kicked off in August 2007 at a meeting in Amman, Jordan. Foster Wheeler sent two experienced pipeline engineers to lead the study, Stephen Ward and Graham Prichard. The oil ministry's South Oil Company (SOC) sent a delegation led by Kifah Numan, the deputy director general at SOC. Commander Angel Salinas and I represented GRD at the initial meeting, as well as at follow-up meetings over the next three months.

SOC shared specific data of their current operations as well as the objectives of any SPM export project. Specifically, SOC targeted the capability of exports of 4.5 million barrels per day out of the south. This was three times what their export volumes were at the time. Foster Wheeler took the data and objectives back to its offices in the United Kingdom.

In early November 2007, I received a phone call from one of the Foster Wheeler pipeline engineers assigned to the study. "Gary, are you sure that the information you provided on the two subsea forty-eight-inch pipelines going to ABOT is correct?" he asked.

The only information that we had on those two pipelines was that the last time they had been inspected was 1989, just before the first Gulf War. The inspection had been accomplished with an ultrasonic pipeline inspection gauge (PIG) to measure the thickness of the pipe from shore to ABOT, more than thirty miles. The inspection had revealed severe internal corrosion. It was recommended that the pipeline be derated and that the maximum operating pressure decrease from 680 psia to 180 psia. That was all the information that was shared by the SOC leadership. I also confirmed the same information from the contractor who completed the inspection in 1989.

"Why do you question that information?" I asked.

"Based on your input and my calculations, the pipe should have rusted all the way through the steel in the late 1990s," he said.

This was news to anyone that had worked with Iraq's oil sector from the US government. We knew that the subsea pipelines were old and poorly maintained. We just did not know how serious the threat was.

"We will be with the oil ministry's South Oil Company leadership in your offices in two weeks. Discussion of this information will be a high priority," I said.

Two weeks later, we assembled at the Foster Wheeler headquarters in Reading, UK, to review the findings and recommendations of the feasibility study. Before the meeting started, I pulled the four senior leaders from SOC together: Kifah Numan (deputy director general), Asri Moussa (planning manager and previous deputy director general before 2003), Oday Al Quoraishi (project manager), and Saad Joseph Mohammed (deputy project manager and previous terminal manager of ABOT).

I shared with them the comments made by Foster Wheeler that the existing forty-eight-inch pipelines supplying crude oil to the offshore export terminal should be leaking profusely based on the information received from these four gentlemen at our August meeting. I asked what they were doing to mitigate this serious threat.

Asri Moussa went first. He used to have nightmares in the late 1990s about the subsea pipelines failing. Moussa was the deputy director general at the time and he knew that when these pipes failed, Saddam would send him directly to prison.

"Are you still having nightmares?" I asked.

Moussa said that he no longer had the nightmares since he did not have operations responsibilities anymore. He looked at Kifah, standing next to him, who currently held the position.

I looked at Kifah and asked him what SOC was doing to mitigate the problem. He said that this subject had been constantly communicated to the oil ministry headquarters in Baghdad, but no one wanted to listen. He was tempted to share this information with the press, just to get the attention of the politicians in Baghdad.

Knowing that Saad had been the terminal manager for ABOT until his recent appointment to this feasibility study, I asked him if there were operational procedures being followed to mitigate the threat of a failure of the existing pipelines.

Saad looked at me and said in a concerned voice that he prayed five times a day for those pipelines.

At this point, I recognized that the problem with the subsea pipelines was extremely serious. I looked at Saad and said, "Please, do not stop praying."

Just then, Deputy Minister Ahmed Al Shamma walked into the meeting room. I motioned for him to join our small group. When he approached, I summarized what the four SOC managers had just said. Shamma mentioned that he had not been able to get the attention of the oil minister about the seriousness of this problem.

"Sir, you need to take this on and convince Minister Shahristani of the problem. Share this with him. I am leaving for Washington, DC, after our meetings in Reading for a short visit. While I am in DC, I plan to share this information about the pipelines with senior people within my government. I will also share this information with the GRD leadership in Baghdad and request they share it with the commanding general and the US ambassador. Let Minister Shahristani know that this problem is serious and cannot be ignored," I said.

The meeting in Reading was successful and the engineers with Foster Wheeler did a nice job of providing a feasibility study. The major conclusion was that Iraq needed to invest about $6 billion in their southern export infrastructure, $4 billion onshore and $2 billion offshore. This was required to meet their target export capacity of 4.5 million barrels per day.

I contacted the most senior person within the Bush Administration that would understand the seriousness of the problem. Reuben Jeffery had been part of the CPA group during the summer of 2003 and had been appointed director of the CPA West group at the Pentagon during the second half of 2003 and 2004. Reuben was now the undersecretary of state for economics in Foggy Bottom. He agreed to see me in early December 2007.

Reuben brought together several staffers at the State Department to sit in on our meeting. I shared with the group everything that we knew and also shared my personal assessment. Our biggest problem was that we did not know if the pipelines would last another day, week, month, or year. If they failed, it would probably be a catastrophic failure. I shared the stories of how the GRD contractors had experienced the shaking of the ABOT offshore terminal because the relief valves were frozen shut during pressure surges.

Reuben and his team were sympathetic and recognized the problem, but they let me know that their hands were tied. Reuben said that Congress would not entertain any more US government funds for the Iraqi oil sector. It was December 2007; the Iraq War was not popular on Capitol Hill.

The message was clear. Iraq would have to fund any export infrastructure project. As a US taxpayer, I agreed with the message. However, I knew that using Iraqi funds would make the project much more difficult to successfully complete.

Deputy Minister Shamma succeeded in Baghdad at convincing Oil Minister Shahristani of the need to address the current export infrastructure problem.

In mid-January 2008, a meeting was held at the oil minister's office inside the Green Zone, close to the Energy Fusion Cell. Mark Tokola and Major General Jim Snyder of the Iraq Transition Assistance Office (ITAO) led the

US team. Captain Tom Brovarone, Commander Phil Malone, and myself represented GRD. Commander Malone had recently replaced Commander Angel Salinas as the government lead for the project.

The oil minister requested US help to address the serious problem with the southern export system. He stated that it was his number-one priority. He asked that the United States provide a bridge between the feasibility study that was funded by the US government and a front end engineering design (FEED) contract with Foster Wheeler that would be funded by Iraq. He wanted the United States to keep the Foster Wheeler engineers engaged and focused on this important project. He needed a bridge period of five months.

The meeting ended with the oil minister asking for a long-term commitment from our group. The ambassador and the general advised him that they would be departing Iraq soon and that the other military people in the room would also be departing within a few months. The oil minister looked at me. I commented that he had my commitment to stay the course. I agreed to remain on the project for as long as it took or as long as the US government would renew my contract, whichever came first. I also commented that my minister of interior (Iraqi slang for wife) would have to agree. DOD honored its commitment and kept me under contract until the departure of the US forces almost four years later.

Over the next couple of months, the oil minister made a decision that would cost his country hundreds of billions of dollars in lost revenue over the next ten years. He decided to divide the project into two separate projects. South Oil Company (SOC) was responsible for the offshore project. The State Company for Oil Projects (SCOP) was responsible for the onshore project.

This decision made little sense to me. SCOP did not have the capacity to manage the $4 billion onshore part of the project. They had been hit harder than any other company under the oil ministry since 2003. Many of their competent employees had departed, had retired, had been killed, or had quit. Their Sunni director general had been kidnapped in 2006 and assumed killed. They had not been able to successfully complete the smallest and easiest projects. On the other hand, SOC had developed a good reputation with the GRD contractors. Their senior management demonstrated competency. Their manager of capital projects, Oday Al Quoraishi, was competent, had excellent English-language skills, and understood Western management practices.

I complained to Deputy Minister Shamma, only to learn that Shamma had argued with the oil minister to give the entire project to SCOP. His argument was that there existed an old law that dictated all major oil infrastructure projects be managed by SCOP. As the executive champion for the entire project, Shamma was more comfortable working with SCOP than SOC. He had spent most of his career in the SCOP organization.

The oil minister felt that neither SCOP nor SOC had the expertise to handle the $2 billion offshore piece of the project. This was the most critical and would require technical expertise that existed only with international companies. Iraq had fought three wars since the last major offshore work had been accomplished in the mid-1970s. The extent of unexploded military ordnance sitting on the ocean floor in the northern gulf was unknown. Assistance would be needed from the coalition navy, as well as from international engineering and construction companies. SOC would manage all the contracts supporting the offshore part of the project. The minister requested that GRD work with SOC to support the offshore piece.

GRD agreed to fund and manage what we called the pre-FEED part of the offshore project. It involved the issuance of a task order to Foster Wheeler under their existing GRD contract to continue with engineering work. Besides engineering, it was required to find competent international companies to perform subsea surveys in the waters of the northern Arabian Gulf.

Commander Phil Malone, USN, was the US government contracting officer. Phil's first challenge was to find $2 million for the task order. He had to go to Congress for the funds since all oil reconstruction funds had already been spent. The United States was spending millions of dollars every day in Iraq, but the fiscal lawyers in Baghdad felt that any additional funds spent on the oil sector had to get congressional review. Phil and Steve Lindsey from the ITAO group did great work to quickly handle the congressional notification.

Foster Wheeler appointed an experienced senior project manager, Malcolm Corry, to manage its team. Corry documented and explained the Foster Wheeler procurement system to the Iraqi government auditors before using it to solicit competitive bids. Two immediate contracts were needed to perform subsea surveys: bathymetric and unexploded ordnance surveys. The bathymetric survey created a detailed map of the ocean floor. The unexploded ordnance survey identified any metal object lying on the ocean floor and up to three meters below the floor. Both contracts were funded by Iraq.

As part of the engineering, Foster Wheeler completed a project schedule that showed SPMs floating in the northern Gulf during the third quarter of 2010, about thirty months in the future. Several assumptions were used to develop that schedule. The key assumptions that proved faulty were that the FEED and subsea survey contracts would be executed by the oil ministry before June 2008 and that all future project contracts would be "fast-tracked" through the approval process at the oil ministry in a week. Neither of these assumptions held true.

The deadline for contracts of June 2008 came and went. Other distractions at the ministry headquarters prevented a real focus on this project, not the least of which was the pressure from the Council of Representatives to seek the removal of the oil minister. Another distraction may have been the increased violence in the south during the Charge of the Knights. This was an Iraqi military operation against the Shia militia in Basra.

Commander Malone put enough contingency in the pre-FEED task order to keep the contractors engaged until the end of September 2008. Phil's tour ended during the summer of 2008 and his replacement was Air Force Major Jarrett Purdue. September also passed without any movement by the ministry headquarters.

The nightmares that Asri Moussa said he experienced in the 1990s were now haunting me. If the existing pipelines failed and Iraq lost half or all of its revenue for a period of several months, what would the impact be on the stability of the country? What would be the impact on the success of the surge? Could the fragile Maliki government survive? What could I do to progress this project?

I often visited my dear friend Thamir Ghadhban to seek his advice. He was not in a position to help at the time. Shahristani saw Ghadhban as a threat and would not listen to him. It would take another year and Thamir saving Shahristani's job before the oil minister would welcome Thamir's advice. I will cover that in my chapter on the oil field bid rounds.

A former oil ministry director general who was working for Ghadhban recommended that I request a private meeting with the oil minister and jump up and down on his desk to get his attention. I recognized his figure of speech, and although it was very tempting to take his advice literally, I stayed in my authorized lane of working with the deputy minister.

Then, Iraq's luck changed. Oil Minister Shahristani was in London for a conference in October 2008 when a reporter with the *Financial Times* asked him about a serious threat to the subsea pipelines. The reporter highlighted that ABOT accounted for 95 percent of Iraq's revenue. What was Shahristani doing to mitigate this very serious threat to his country's security?

Shahristani's response was that the SPM project was being worked on and that there would be SPMs floating in the northern Gulf by the third quarter of 2010. He was quoting the schedule developed by Foster Wheeler earlier in the year. However, since none of the deadlines had been met, that schedule was overcome by events. What was important was that the oil minister was now on public record saying that the problem would be solved.

South Oil Company signed a FEED contract with Foster Wheeler and two subsea survey contractors in early 2009. Serious work started.

John Scott was the manager of projects at Foster Wheeler. John had been in the oil industry for more than twenty years, working for Texaco before joining Foster Wheeler. He recognized the importance of this project and quickly beefed up resources to support Malcolm Corry, the Foster Wheeler project manager. Several experienced engineers were appointed to the team: Joe Emans, Roger Kempston, Graham Pritchard, Richard Large, Tim Herbert, and others. All had excellent experience in the international project space.

The Iraqis kept Oday Al Quoraishi as SOC project manager and Saad Joseph Mohammed as his assistant. Baghdad appointed Adnan Al Gharawi from the Planning Directorate to represent the oil ministry.

The coalition forces assigned the highest priority to the project. The following US military units played significant roles: Corps of Engineers Gulf Region Division (GRD), Coalition Navy, Energy Fusion Cell, Civil Affairs Command, US Division South, and the US Forces Staff. The list of names is endless, but a few that come to mind and have not already been mentioned include Colonel Rick Waddell, USAR; Lieutenant Commander Brian Delaney, USN; Lieutenant Commander Mike Cookson, USNR; Colonel Greg Zehner, USAF; Commander Bobby Quintos, USNR; Major Vern Starman, USAF; Major Jeff Steinbrink, USA; Rear Admiral Mark Belton, USNR; Brigadier General Bob Ranck, USAF; Captain Jeff Mobed, USNR; Commander Mike Tasker, USN; Commander Andy Elvin, RN; Commander Ben Beech, RN; and my colleague in Baghdad Kevin Ross.

Several design meetings and conference calls were held in early 2009. The FEED work progressed quite well. An appropriate project name was agreed on at one of the early meetings. The new name was the Iraq Crude Oil Export Expansion (ICOEE) project.

In June 2009, a meeting was held at the Foster Wheeler offices in the United Kingdom. It was the first of many interface meetings with SCOP's onshore project team. I had previously challenged the wisdom of awarding such a strategic project to the SCOP organization, one of the weakest organizations under the oil ministry. I was curious about their progress. I and representatives from SOC, SCOP, Ministry Planning, Foster Wheeler, and the coalition navy were in attendance.

The morning meeting went well, with Foster Wheeler sharing the offshore design. Then, the first indications of serious problems with SCOP's onshore piece became evident after lunch. Rebuilding storage tanks at the Al Fao terminal was well behind schedule, and the project manager gave no one in the room confidence that he had a plan to fix the problems. Then we arrived at a significant interface point between the two projects. SCOP was responsible for procuring and installing the onshore pumps. The sizing of the pumps was of critical importance to the success of the ICOEE project.

Foster Wheeler determined that the offshore system would require ten large gas turbine pumps at the Al Fao terminal. These pumps were necessary in order to operate the SPMs at their maximum capacity of nine hundred thousand barrels per day. Such large pumps had a two-year procurement lead time. The current ICOEE schedule anticipated a fourth quarter 2011 start-up of SPMs, less than thirty months away. Procuring and installing the pumps was the responsibility of SCOP. The message to the senior SCOP managers in attendance was that they needed to place the pumps on order as soon as possible.

The senior SCOP manager objected to the size of the required pumps. Foster Wheeler made a detailed explanation as to why the larger pumps were required.

The senior design manager from SCOP refused to accept the design requirements. He only said that he would have to review with his director general in Baghdad. I later learned that SCOP had already ordered smaller pumps. They ordered the same-sized pumps that had been initially designed for Al Fao back in the 1970s.

This was the first interface meeting with SCOP. It highlighted my concerns about the oil minister's initial decision to have SCOP manage the onshore work. They truly lacked the capacity to handle such a strategic project for their country. Ordering the large gas turbine pumps for Al Fao became a point of discussion for all future meetings held with the SCOP leadership in Baghdad. Ordering the gas turbine pumps was always on the discussion agenda whenever the US ambassador met with the oil minister during 2010 and 2011.

My last project interface meeting with SCOP was in September 2011. SCOP still had not placed the order for the gas turbine pumps. A press interview in 2015 with a deputy minister mentioned that the pumps might be installed in 2018. Fortunately, workarounds that use pumps located close to the oil fields have enabled the SPMs to load at half-capacity, but the SPMs will not be able to reach their optimal export levels until the gas turbine pumps get installed at Al Fao.

Another memorable project meeting in 2009 was about determining if Iraq should risk an inspection of the old pipelines with an ultrasonic pipeline inspection gauge (PIG). These forty-eight-inch subsea pipelines had last been inspected in 1989 and were found to have severe internal corrosion. Deputy Minister Shamma led the group in a meeting at Foster Wheeler with a pipeline inspection company. The expert from the company was well known in the industry and had worked in Iraq in the past.

He made an excellent presentation highlighting how his company would ensure that the inspection PIG would not get stuck in the pipe or create a pressure surge that might cause the pipeline to fail. Deputy Minister

Shamma asked many questions to better understand the risks to the pipeline, if he approved such an inspection. The salesman offered many verbal guarantees that his company would succeed in performing the inspection without damaging the pipelines. Shamma kept challenging him about the verbal guarantees. After about the third question, I asked the deputy minister if I could ask his question in a different way. Shamma smiled and encouraged me to ask the question.

I highlighted the tremendous risk that the salesman was asking of the deputy minister and the oil minister. His verbal guarantees were good, but the best way to get the oil minister comfortable with such a decision would be for his CEO to guarantee that his company would cover all lost revenue to Iraq if these inspection services failed to deliver as promised. The salesman quickly responded that he could never do that. Deputy Minister Shamma looked at me and said that the response provided the clarity he needed.

The bottom line was that Deputy Minister Shamma recognized that there was very little upside benefit in conducting an inspection on a pipeline that was currently providing almost all of his country's revenue. However, the downside was significant. The two old subsea pipelines were not inspected.

The meeting also provided an opportunity for an excellent discussion about why the subsea pipelines had not yet failed. Everyone in the room agreed that there were probably spots along the pipe that had rusted all the way through the steel. So, why was it not leaking?

Deputy Minister Shamma was a project manager on the original project in the early 1970s when Brown & Root installed the pipelines. He confirmed that the pipelines were supposed to be buried by as much as three meters in the seabed. The seabed in the northern Gulf was almost all silt that was deposited at the end of the Tigris and Euphrates Rivers. This silt in twenty to thirty meters of water would become a compact mud around the pipelines over years of settling. He also confirmed that the pipe had a concrete coating to protect the exterior.

We speculated that the severe corrosion detected in 1989 was more than likely on the bottom of the pipe in low sections of its route. We assumed that most of this corrosion took place during the Iran-Iraq War. The pipe was shut down over a period of several years during that war. The water in the oil settled to the bottom of the pipe and started to oxidize with the steel. It was felt that the normal buildup of sludge in the pipeline would also build up in the same low spots along the route where the corrosion took place. The combination of the sludge inside the pipe, the concrete coating on the outside, and being encased in mud under pressure from the sea had prevented any serious blowout of the pipe. This was our speculation. It made sense to everyone in the room.

Subsea pipelines normally start failing at the waterline, where the pipes exit the water. There were three such leaks with the forty-eight-inch pipelines in 2009: two at ABOT and one at the shore. Repair of these waterline leaks was relatively easy. The US Navy assisted SOC and quickly installed clamps on the two leaks at ABOT. SOC repaired the leak at shore. Downtime and environmental damage were minimal.

The project transitioned from the design phase to the execution phase in 2010. Long lead materials were placed on order. They included pipe joints, subsea valves, SPMs, and the like. Major contracts were negotiated with Foster Wheeler for the project management contract and eventually with Leighton Offshore for the phase one construction contract.

Many of the procurement actions for long lead materials were delayed due to the slow contract approval process at the oil ministry. A couple of procurement actions had been delayed for as long as two months. I attempted to hurry things along with the project-critical path material by identifying the opportunity costs to Iraq for these delays.

In 2010, international oil companies had submitted their production plans for increasing production in the giant Rumaila, Zubair, and West Qurna oil fields. Their plans indicated that production would surpass existing export infrastructure in the fourth quarter of 2011. Production would be more than five hundred thousand barrels per day higher than the current export infrastructure could handle. Any delay to the ICOEE project would mean that those barrels could not be produced. It would be capped production. The opportunity costs would be $37 million per day, assuming an average market price of $75 per barrel.

At a meeting in early 2010 in Deputy Minister Al Shamma's office, I emphasized that a specific contract sitting in his inbox at the time was costing Iraq $37 million per day for every day that it sat without approval. "Sir, what else is the ministry doing that is more important than processing the contract in your inbox?" I asked.

Deputy Minister Shamma replied that I was not being fair to him.

"Sir, these delays are pushing the start-up of the ICOEE project into 2012. It truly is $37 million per day at the current market price. Why am I not being fair?" I asked.

He then shared with me what I had suspected. Iraqi government managers had been sent to prison for not following detailed government procurement regulations to the letter, or for just making honest mistakes. The threat of a prison sentence was driving behavior, poor behavior. He said that no one in Iraq had been sent to prison for losing billions of dollars in opportunity costs, but many had been sent to prison for being suspected of losing thousands of dollars on a contract.

"Fair enough, but please sir, this opportunity cost is serious. I am going to start sharing this number with my leaders at the embassy, so I suspect the ambassador will be quoting that number with the oil minister and the prime minister," I replied.

A project management contract with Foster Wheeler was executed in mid-2010. A new manager was selected, Malcolm Shepherd. Malcolm had just completed managing a large project in Asia. Also, Manaf Al Jadir was brought in for this contract. Both managers had excellent credentials and experience in the international project management space. Manaf offered additional expertise. He was born and raised in Iraq, the son of a former executive in the Iraq Petroleum Company (IPC). Manaf had just completed a large project in Saudi Arabia. Some of the Foster Wheeler design engineers moved onto the new contract; the others remained to close out the FEED contract.

Another important interface meeting held in 2010 was with the Japanese Sealine project team. This included representatives from Japan International Cooperation Agency (JICA) and their contractor, Japan Oil Engineering (JOE). It was in Amman, Jordan, and was the second interface meeting between the ICOEE and Sealine teams. The first had been held in Istanbul several months earlier. The quality of the questions asked by JOE during the first meeting was poor, so I did some research to better understand their experience with large offshore oil pipeline projects. This offshore project was the largest project that JOE had ever undertaken and it was significantly bigger than any of its other projects.

The Amman meeting was more troubling than the first meeting in Istanbul. JOE shared their project design with the engineers from SOC and Foster Wheeler. It was alarming. They were introducing risks not only to their project but also to the ICOEE project because of their proposed pipeline route.

Oday Al Quoraishi quickly pointed out to them that their design introduced huge risks and would not be acceptable to SOC. The meeting was conducted in English and Oday's words were direct and to the point, but the JICA representative must have felt offended. Her response surprised me: "SOC is not our customer. This is a government-to-government agreement. Our customer is the Ministry of Oil and Thamir Ghadhban in PMAC."

Oday stood up and left the meeting. He later told me that he was frustrated with the attitude of both JICA and JOE. He understood that the low-interest loan was on a government-to-government basis. That was not his problem. He was frustrated because there were obvious safety issues in their design and SOC must be the customer for anything related to safety. He was right. SOC was the operator and safety was its responsibility.

During Oday's absence from the meeting, Manaf Al Jadir from Foster Wheeler walked the JICA representative through the reasons why its proposed design was unacceptable to SOC. He did a nice job explaining the flaws in its design and proposed a different pipeline route that kept its pipe far from the ICOEE pipeline.

I approached the young JICA representative at the break. She was still adamant that its design had just been reviewed with Thamir Ghadhban and it could not be changed. I assured her that Ghadhban would never overrule design safety objections that were just made in our meeting.

JOE eventually developed a new pipe routing design that was acceptable to SOC. Its new design kept its pipe and valve stations away from the ICOEE pipes. However, JOE continued to have difficulties accepting the customer relationship with SOC. This contributed to project delays. As of 2016, the project was 90 percent complete. No progress was being made due to technical disagreements.

The ICOEE phase one construction contract was awarded to Leighton Offshore of Australia. Several representatives from Leighton, Foster Wheeler, the oil ministry, and SOC met in Basra five days before signing the final contract. The formal signing would be at a public ceremony with the Basra governor and upstream Deputy Oil Minister Karim Luaibi. Luaibi had taken over the role of project champion at the ministry during the summer of 2010, and he would soon be appointed to the oil minister position, in 2011.

Malcolm Sheppard from Foster Wheeler approached me at the end of the second day and advised that Leighton was leaving. Malcolm mentioned that the oil ministry had made some late changes to the contract terms and conditions that were unacceptable to Leighton.

I immediately spoke to Oday Al Quoraishi and asked why the late changes to a contract that had been in play for the last nine months. Oday was just as frustrated as the rest of us. His hands were tied. The changes came from the oil ministry audit and legal departments. His perception was that the changes were government-mandated.

We did not have time to fool with an additional delay. If Leighton walked away at this point, it would probably mean a delay of several months to the project, at a minimum. I approached the two senior Leighton managers at the hotel where we were all staying and asked what the problem was. The Leighton executive told me that the oil ministry had made significant changes to the original terms and conditions that were agreed to when the proposal was submitted. Additional negotiations in Istanbul the previous month did not reveal new terms and conditions. He was certain that his CEO would not accept the new risks.

They were packing their bags and getting ready to leave the hotel. I asked them to allow me to show them something. I shared a PowerPoint slide presentation that my colleague Kevin Ross had prepared in Baghdad. It was prepared for the US embassy using public information, but Kevin made a couple of slides that did an excellent job of portraying the enormous Iraqi potential for crude oil production over the next ten years. I explained to the Leighton executives that if they walked out now, they would run the risk of irritating the ministry leadership and being blacklisted from future contracts. The names of the oil companies and the size of their future investment in Iraq caught their attention. They decided to stay another day. One request was made of me: "Gary, please show those slides to our CEO when he flies in tomorrow."

The Leighton CEO arrived the next day on his private jet and came to the hotel. I went through the same slides with him. The CEO recognized the risk he would take if they walked away from this large contract that was so important to Iraq's future. He was not pleased with the late changes to the contract terms but decided to stay for the signing ceremonies. He knew he would have to convince his board of directors back in Australia of his decision. He looked at me and asked, "Gary, can I get a copy of those slides you just showed me so that I can share them with our board of directors?"

The signing ceremony was a public relations success. It was held at the new SOC training facility in Basra, a very nice facility with a four-hundred-seat auditorium. It was well decorated and there were close to a hundred members of the Iraqi and international press. The Basra governor, the upstream deputy oil minister, SOC's director general (Dheyaa Jafaar), Leighton's CEO, and Malcolm Shepherd of Foster Wheeler gave speeches. Manaf Al Jadir sat next to me and helped me with the translation of the Arabic speeches. The $750 million contract signing took place onstage with full press coverage. The Leighton contract was the last major contract needed to complete phase one.

Phase one of the ICOEE project started up in March 2012. This phase of the project consisted of two new forty-eight-inch subsea pipelines from the beach to an area southwest of ABOT. Three other forty-eight-inch pipelines were installed from the beach to a location about eight kilometers from shore, beyond the tidal mudflats. These pipelines were for future export projects. A large subsea valve station was installed on the ocean floor close to ABOT. Three SPMs were installed to the southwest of ABOT on the southern side

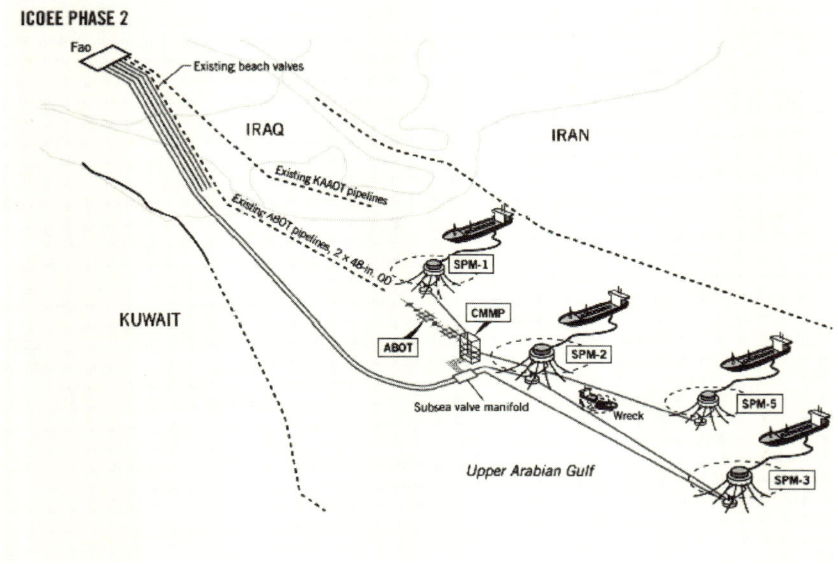

Drawing of the export project off the shore of Iraq in the North Arabian Gulf. Drawing by Foster Wheeler.

of the Khawr Al Kafka channel, the deepest channel in the northern Gulf. A fourth SPM was installed on the east side of ABOT in the same channel. A pipeline manifold was installed on-shore at the Al Fao terminal dedicated to the SPM export system.

Initially, SOC developed a workaround to pump the oil from a pump station in the south Rumaila oil field through the Al Fao manifold to the SPMs. Other workarounds have been implemented to pump more oil to the SPMs, but the SPMs remain operating at a level significantly below their capacity of nine hundred thousand barrels per day.

Phase two of the ICOEE project started up in 2014. This phase consisted of a large central metering and manifold platform (CMMP) installed by SAIPEM of Italy. Only two SPMs could operate under phase one. The CMMP enabled all four of the SPMs to operate. However, one SPM was still not operating at the end of 2015 and awaited the removal of a wreck. The VLCC Amuriyah sank in the channel during the first Gulf War in 1991 and had never been removed. It sits in the waters close to SPM-5, prohibiting clear navigation.[2]

The ICOEE project was an important part of my life for five years. The US taxpayer invested $2 million to seed a project that cost the Iraqi government $2 billion. The project will deliver hundreds of billions of dollars in

The central metering and manifold platform installed at the south end of the Al Basra oil terminal. Photo by Foster Wheeler.

revenue to Iraq over its lifetime. I am proud to have worked with a team of professionals from the private sector, South Oil Company, the Ministry of Oil, and especially the US forces to help overcome many hurdles and deliver on this important national security project for Iraq.

An important note should be mentioned here. A few months after my departure from Iraq, this project became tainted by a bribery investigation of Leighton Holdings by the Australian authorities. The results of this investigation have not yet been made public. It pains me that such a wonderful project for Iraq should be tainted by a corruption rumor. Leighton was also the prime contractor of the JICA Sealine project.

CHAPTER 18

The Bid Rounds

Oil Field Auctions

The negotiations have only just begun.
—BP oil executive

Iraq signed contracts with some of the world's largest oil companies in 2009 and 2010 to develop Iraqi oil fields and produce the incremental production of several million barrels per day. These contracts obligated oil companies to increase Iraq's oil production by more than eleven million barrels per day before 2018. Such production levels would cause Iraq to surpass Saudi Arabia as the number-one producer and exporter within OPEC. Contracts with these lofty production targets had never been seen within the global energy sector. These contracts could be the impetus to push Iraq into a new era. The process that enabled these contracts was known as the bid rounds.

Iraq conducted two bid rounds in 2009: one on June 30 and one on December 12. Two additional bid rounds took place later: bid round number three in 2010 and bid round number four in 2012. It is the first two that could potentially propel Iraq forward as a global energy leader.

In hindsight, these bid rounds were eventually successful for Iraq and the Iraqi oil sector. This was not always the case, especially just after the first bid round. The look on Oil Minister Shahristani's face at the end of the session on June 30 was one of defeat. It was a troubled look, the look of failure.

How did Minister Shahristani and the Maliki government capture victory from the jaws of defeat? What really happened following the first bid round? How did things play out over the course of the execution of the contracts? What would prompt a senior executive of one of the largest oil companies to regret ever signing a contract? It is important to start with the political pressures on Minister Shahristani in 2008.

It was early March 2008. Dathar Khashab phoned Oil Minister Shahristani to invite him to attend a small refinery-commissioning ceremony in Najef scheduled for the upcoming weekend. Dathar was the successful director general of the Midlands Refining Company in Baghdad. He was starting up a new crude oil unit at one of his small refineries close to the city of Najef. It was the first capital project start-up in the oil sector since the oil minister's appointment two years earlier. Shahristani initially declined, stating only that he had too many problems in Baghdad with parliament to be spending a day in Najef. Several members of parliament were calling for his resignation because the oil sector had been stagnant under Shahristani's leadership. Dathar persisted until Shahristani agreed to attend. But Shahristani only agreed to a half-day.

Dathar attended ceremonies in his signature blue refinery coveralls. He had worked hard to get this new crude unit project started and completed, in spite of all the delays and roadblocks by the oil ministry headquarters. The contract procedures that the oil ministry imposed on Dathar were stifling. He frequently said to me that it was easier to do business with Western companies during the sanctions regime than under the new and enlightened "American contract process." Dathar gave all the credit to Shahristani at the ceremony for this crude unit start-up. The celebration was a huge success and dignitaries from throughout the Najef Shia religious community attended. There were more than a thousand people in attendance; many were not just religious leaders but also political leaders from the Shia religious political parties. Shahristani used the opportunity for his political survival. He spent two full days in Najef and then went to the city of Diwaniyah, giving speeches and making promises of great things to come from the oil ministry, if he could get the support of the powerful Shia political parties. It worked. Shahristani's political career was probably saved that weekend, thanks to his senior Sunni director general, Dathar Khashab.

Oil production had been in decline since Shahristani's appointment in May 2006. The US government funded and managed a well work-over project in late 2006 and early 2007 that slowed the decline significantly, but the deterioration became evident to everyone in early 2008. Exports in the south were averaging less than 1.5 million barrels per day. Revenues into the country's treasury were affected. Shahristani could not come up with a plan to stop the decline and reverse the trend. All of his operating companies, except for Dathar's refining company, were failing to spend the budgeted capital investment required to rebuild the oil sector.

There were many reasons for the oil sector investment failure. South Oil Company was the largest company under the Ministry of Oil. It produced most of the country's oil from the supergiant oil fields near Basra. They had difficulties engaging international oil service companies to sign drilling contracts for the southern oil fields. Without the necessary drilling rigs, the southern production declined. The problem was not an issue of cost. It was more of a question of procurement capacity, guidance, and restrictions on South Oil Company and other oil sector companies' ability to even talk with service companies. The review of contracts by auditors and the inspector general also slowed the process to glacial speed. The history and ability of the inspector general to imprison oil sector leadership for suspected contract violations hampered progress. The new oil minister did not have either the ability to fix the problem or the political will to tackle such problems, which were shrouded with clouds of corruption.

With renewed support from the Najef Shia leadership and Prime Minister Maliki, Shahristani had new life, but there was a time limit. He had to come up with a better plan. He had been negotiating directly with the largest international oil companies in the world to enter into no-bid contracts. These contracts would have been under the umbrella of the new oil laws, which had been sitting in parliament for more than a year. Those same members of parliament who called for the minister's resignation were the biggest reason why the petroleum and revenue-sharing laws were stuck in parliament.

Shahristani announced a change in direction at the end of June 2008. Those same large international oil companies that had been negotiating for no-bid consulting contracts would be invited to bid in 2009 on developing the largest oil fields in Iraq. The oil ministry's Petroleum Contracts & Licensing Directorate (PCLD) received the mission to execute the plan. Natiq Al-Bayati was the very capable director of PCLD. Natiq had worked in the upstream sector of the ministry for more than thirty years and was highly respected by his peers. However, such work would require more expertise than the oil ministry could muster from within their ranks. An international consulting firm, Gaffney, Cline & Associates, was contracted to assist PCLD.

Shahristani made a bold move by announcing this new direction. It would help rebuild the oil sector and save his political career. However, there would be more critics from parliament and even more from within his own oil ministry.

The first challenges confronting Shahristani were the legacy production-sharing contracts signed by Saddam during the 1990s. The Maliki government had to address those contracts before proceeding with any type of oil field development contracts. The production-sharing contracts had been negotiated with oil companies from countries that were members of the UN Security Council in an attempt to solicit their support for removing UN sanctions prior to 2003. The two most significant contracts that had been signed were with the Russian company Lukoil for the West Qurna oil field and with the Chinese company CNPC for the Ahdab oil field. The Lukoil contract had been canceled in a controversial move by the Iraqi parliament in December 2002 for failure to perform. The post-2003 oil ministry's legal department took the position that Iraq had legally canceled that specific contract and therefore had no legal liabilities to honor it. The other contract was with CNPC. It was still valid and needed to be addressed. That contract had been signed in 1996 between CNPC and Iraq. No work under the contract was accomplished during sanctions, but Iraq did nothing to cancel that contract as it did with Lukoil. Shahristani negotiated a technical services contract with the Chinese in November 2008 to replace the old production-sharing agreement. This technical services contract would be the first of many future oil field development contracts.

PCLD and its consulting firm conducted workshops and maintained communications with the global energy community to share the necessary information with the oil companies. This work helped the oil companies prepare bids for the designated oil and gas fields in the bid round in 2009. A date was set for the last two days of June 2009 in the Al Rashid Hotel inside the Baghdad Green Zone.

Shahristani's critics intensified their criticism as the date grew closer. His critics in parliament were loudest just days before the bid round date. Shahristani's biggest criticism was coming from the Kurds, who wanted to sign their own oil contracts with foreign firms. Criticism came from the oil unions, Iraqi oil industry experts, Iraqi citizens, and others that wanted only a limited presence of foreign oil companies in Iraq. The leadership of South Oil Company was among his biggest critics.

The first bid round included the largest oil fields, which South Oil Company had been operating for decades. Specifically, Rumaila, Zubair, and West Qurna were part of the first bid round. South Oil Company criticized any plan that brought in foreign oil companies to develop those large oil fields already being managed by South Oil Company. It was difficult for the South Oil Company leadership to accept that their oil minister was essentially telling them that they needed outside help. Shahristani understood the criticism, but South Oil Company could not reverse the production decline of the last few years and Iraq needed that reversal to occur quickly. He and Iraq could wait no longer.

South Oil Company had four different directors general within a year before the first bid round. The very capable Jabbar Al Luaibi had been the director general since April 2003 and had been promoted into the position of advisor to the oil minister for the southern oil sector. Kifah Numan took over the position in early 2009 and was ordered to open the valves on the production wells to increase production. Such action would provide increased short-term production but substantially increase the long-term risk of damaging those producing wells. Kifah was an experienced oil technocrat that was highly respected within South Oil Company and the international oil community. However, he was not politically aligned with any party within the Basra community and had no desire to become aligned. Recognizing the realities of his political status and the long-term damage that opening the valves could do to the oil fields, he resigned his position.

The director general of the Planning Directorate, Fayadh Nema, was sent to Basra as the new director general in the spring of 2009. Fayadh had never worked in the upstream portion of the oil sector. Most of his career had been in the Ministry of Industries; in 2004, he joined the oil ministry in a middle-management position. Fayadh was well liked within the ministry headquarters and throughout the Maliki government. He was Sayyid, a title denoting males accepted as descendants of the Prophet Mohammed. Fayadh's decision to temporarily increase production by opening the valves was controversial within South Oil Company, but his most controversial decision was to publicly criticize the upcoming bid rounds at a conference in Egypt. He said that the contracts in the first bid round would put the Iraqi economy in chains and shackle its independence for the next twenty years. This public criticism cost him his job. Prime Minister Maliki and Oil Minister Shahristani removed him from the position he had only held for a couple of months. He was appointed as director general of the Technical Directorate in Baghdad.

The new director general was Dheyaa Jafaar, a Dawa Party member who had also held key positions within South Oil Company for a number of years. Dawa was a Shia religious political party headed by Prime Minister Maliki. Dheyaa was aligned with the powerful Dawa Party in Basra and he was a highly respected production manager in South Oil Company. He would bring some stability to the position until his promotion to deputy minister in 2015. Dheyaa was attending an ICOEE project review with me in the United Kingdom when the decision was made to promote him into the job.

The First Bid Round—June 30, 2009

There was a massive dust storm at the end of June 2009, closing the Baghdad Airport until late on the twenty-ninth. The first day of the bid rounds was postponed until the thirtieth. The two-day event was compressed into one day. All bids for oil fields took place on the thirtieth.

Everyone assembled around 9 AM in a large meeting room at Baghdad's Al Rashid Hotel inside the Green Zone. In the front of the room were dignitaries, including Prime Minister Maliki, oil minister Shahristani, Thamir Ghadhban, and others. There were more than a hundred executives from international oil companies. Several representatives from embassies were present, including the US embassy.

A transparent glass box sat on a table in the front of the room. This was the bid box. All bids were to be sealed and placed inside this box. The first oil field to be auctioned was the supergiant Rumaila oil field, located just west of Basra. The southern tip of this field actually crosses the Kuwaiti border into northern Kuwait. With reserves of seventeen billion barrels, this huge oil field is considered one of the largest oil fields in the world. It was producing about one million barrels per day in 2009.

The clock started for bidders to place their envelopes in the glass box to bid on a twenty-year technical service contract for the Rumaila oil field. The bid had to consist of two numbers. The first number was the amount that the oil company charged for each barrel of incremental production they could produce above a current base production number. This figure was known as the remuneration fee, or the profit for the oil company. The second number was the plateau production target. This was the amount of barrels per day that the oil company would commit to achieving within a six-year period. The winner of the field would be determined by a calculation of those two numbers: remuneration fee and plateau production target. The calculation was simply the plateau production target multiplied by the number of $50 minus the remuneration fee. The oil ministry learned that it was easy to game this formula by just inflating production targets. A change would be made in the second bid round to help correct some of this gaming.

Two consortia submitted bids for the Rumaila oil field: the consortium of ExxonMobil and Petronas of Malaysia, as well as the consortium of BP and CNPC of China. No other bids were placed inside the glass box. Natiq Al-Bayati of PCLD opened both bids and read the numbers out loud to the room. The consortium of ExxonMobil and Petronas had submitted the winning bid of $4.80 with a plateau production target of 3.1 million barrels per day. The bid submitted by BP and CNPC was $3.99 with a production target of 2.85 million barrels per day. ExxonMobil's bid was the better bid according to the formula.

Then, Natiq pulled out a red envelope. He held the red envelope for everyone to see and stated that this was the maximum remuneration fee that Iraq would pay for the Rumaila oil field. Before opening it, he said that even he did not know the contents of the envelope. It became clear to everyone in the room that Minister Shahristani was the owner of this maximum remuneration fee. Natiq opened the envelope and announced the number to be $2.00 per barrel. Jaws dropped in the room. No one could believe this number. Natiq announced it again: $2.00 per barrel. A member of the PCLD later said that no one in PCLD was privy to the contents of the red envelopes. PCLD had recommended maximum fees to the oil minister and the maximum fee for the Rumaila oil field was $3.50. Shahristani had taken the recommendation of his experts in PCLD and slashed it by over 40 percent to $2.00. Why? It made no sense to the oilmen in the room.[1]

Since both bids were above the maximum remuneration fee, neither company qualified as the winning bidder. ExxonMobil had submitted the best bid up until that point, so they were asked if they could lower their remuneration fee to the $2.00 ceiling. They were given several minutes to respond. After a lengthy delay and discussions with Petronas (the Malaysian oil company), they declined. The same question was asked of the consortium of BP and CNPC. The two had been in discussion while ExxonMobil made their decision. So they did not need as much time to respond. They accepted the maximum remuneration fee of $2.00 per barrel, half of their submitted bid. All of the international oil companies were surprised by the BP acceptance. They felt that the fee seemed much too low to get an acceptable return on investment for this oil field.

Natiq moved on to the next field for auction. It was the Mansuriya gas field located in Diyala Province, northeast of Baghdad. No bidders submitted for this field.

The next field was the Bai Hassan oil field near Kirkuk. The consortium of Conoco Phillips, CNOOC, and Sinopec submitted a bid of $26.70 per barrel for the remuneration fee and 390,000 barrels per day for the plateau production. It was the only bid submitted. Natiq announced the maximum remuneration fee for this field was $4.00 per barrel, almost $23 per barrel less than the submitted bid. Conoco declined to lower their fee, so there was no acceptable bid.

The next field was the Zubair oil field, located just to the west of Basra. Like the Rumaila field, South Oil Company had been producing from it since the early 1950s. The production in 2009 was more than one hundred thousand barrels per day. Zubair had 4.5 billion barrels of proven reserves. Several consortia submitted bids for this field. The winning bid was a consortium of ENI of Italy, Occidental of the United States, Kogas of Korea, and Sinopec of China. Their bid was $4.80 for the remuneration fee and a production target of 1.125 million barrels per day. Natiq announced that the maximum remuneration fee acceptable for the Zubair oil field was $2.00 per barrel. Again, no one accepted the maximum fee so there was no acceptable bid.

The rest of the day went the same as the last two oil fields. The remaining fields included the Kirkuk oil field, the Maysan oil fields, the West Qurna 1 oil field, and the Akkas gas field. None was successful. The maximum remuneration fee was less than half the submitted bids. The spread was huge. Shahristani grimaced on several occasions during the announcements of the difference between his maximum remuneration fee and the offered fee. How could there be such a difference between the two? Why was his number so much lower? Many in the room seemed to think that he was intentionally killing his own bid round. Representatives from one of the smaller national oil companies were seen laughing at him by the end of the day. He had placed eight oil fields up for auction and only one had been successfully agreed to.

The look on his face as people walked out of the room was the look of failure. He had been given the opportunity to reverse the production declines in Iraq, but he had failed. Would this be enough for his critics in parliament to render a no-confidence vote and remove him?

In an effort to salvage the day, Shahristani invited some of the bidders who had rejected his maximum fee to submit new bids within twenty-four hours. He said that the Council of Ministers could consider their new bids at their cabinet meeting the following day.

As the executives from the oil companies walked out of the room, several were mumbling about how BP could accept such a low remuneration fee. This low fee could not provide an acceptable return on investment to any of the other bidders in the room. Why did BP break ranks and accept the fee? Several heard the senior BP executive comment, "The negotiations have only just begun."

The next few weeks demonstrated Shahristani's political prowess. He was able to publicly spin the first bid round as being a success. He emphasized the transparency of the process used to auction off Iraq's oil fields. He was able to boast to the Iraqi people that he did not surrender to the large international oil companies. He had protected Iraq's treasure from the greedy oil companies. When challenged about getting only one successful bid, the response was that the volumes from the one field were more than any of the others. He landed the largest field in Iraq.

The reality was different. The winner of the bid round field was supposed to sign a contract within days after the auction. The auction contract had been shared with all the bidders several months prior. The only things missing from the contract were the commercial terms of the remuneration fee and the production target. Weeks went by and there was no signature on the Rumaila contract. The investment community started to question BP's successful bid. It was estimated that such a low remuneration fee could only deliver 5 to 7 percent return on investment. How could BP accept an upstream deal with such low return on investment?

An energy industry consultant met with me while I was back in the United States in the middle of July following the bid rounds. He mentioned that he had just returned from Saudi Arabia on a consulting engagement. He also mentioned to me that the Saudi senior executives within Aramco were celebrating Iraq's apparent failure during their first auction. I inquired as to why they would be so happy. He replied that Saudi Arabia considered Iraq as the biggest threat to its dominance within OPEC over the next twenty years. The failure of the first bid round indicated that its competitor would be delayed an extra year or two. With his permission, I shared his remarks in an email to Thamir Ghadhban. Thamir immediately responded that my timing was excellent. He needed to share my story with the Council of Ministers. Ghadhban never told me why my story was so important, and it would take me a year to learn. An anonymous senior Iraqi colleague of Ghadhban told me the complete story while waiting at the Baghdad Airport.

BP refused to sign the auction contract and was negotiating to improve the economics of the contract. The problem was that Shahristani did not have the authority to engage in one-on-one negotiations for a deal with less economic value to Iraq. The Council of Ministers (COM) had the authority, but they had to be convinced to save Shahristani. The COM eventually decided to appoint a committee to act as its representative in negotiating with BP.

The investment community continued to question the BP management about signing a contract with Iraq that had such a low return on investment—close to 5 percent. Negotiations continued until late September 2009, when Tony Hayward, CEO of BP, announced that the contract with Iraq would soon be signed. He boasted of a return on investment of about 20 percent. He said that Rumaila would take investments of somewhere between $10 and $20 billion to meet the contractual production target. His key comment was that cost recovery would be immediate, so large average cash outlays were not necessary. BP had successfully negotiated better payment terms on their cash outlay.

Tony Hayward would become well known in the United States six months later, after the disastrous Deepwater Horizon oil spill in the Gulf of Mexico on April 20, 2010.

The technical service contracts required that the oil company initially spend their own cash and then be reimbursed by Iraq for those costs. The first half of the twenty-year contract required an enormous amount of average cash outlay to fund the oil field development. The return-on-investment calculation was the profit (or remuneration fee) divided by the oil company's average cash outlays. The remuneration fee was the oil company's payment for their services, and it was fixed. Iraq refused to negotiate a higher remuneration fee. However, the average cash outlay was negotiated to enable the improved economics of 20 percent. It was brilliant. It was a classic win-win. The Iraqi political leadership could continue to boast to the country that they were getting assistance from the oil companies at very low remuneration fees, while the oil companies could tell their investors that they were getting an acceptable rate of return.

After BP announced an expected 20 percent rate of return on the Rumaila oil field, the other participants in the first bid round wanted to participate as well. They all recognized that something significant had changed their original assumptions. With immediate cost recovery, the net present-value calculation for the oil company cash outlays improved significantly. Iraq negotiated similar deals for Zubair, West Qurna 1, and eventually the Maysan oil fields. The consortium of ENI, Occidental, and Kogas came to terms for the Zubair oil field with a remuneration fee of $2.00 per barrel. The consortium of ExxonMobil and Shell agreed to terms for the West Qurna 1 oil field with a remuneration fee of $1.90 per barrel. Contracts for Rumaila, Zubair, and West Qurna 1 were agreed to in November and early December 2009. The contract for the Maysan oil fields was signed with the Chinese later in 2010. Iraq successfully executed contracts on four of the eight oil fields offered during the first bid round. The four contracts called for production from those fields of just under seven million barrels a day within six years. The first bid round was a success.

Another positive result of the first bid round was the improved relationship between Shahristani and Ghadhban. Shahristani no longer looked at Ghadhban as a threat to his job. Shahristani recognized that Ghadhban was a patriotic national asset willing to help anyone who had the best interests of Iraq as their priority. Shahristani invited Ghadhban to attend the monthly oil ministry leadership meetings and looked to Ghadhban's oil team at the Prime Minister's Advisory Commission (PMAC) for assistance during the execution of the bid round contracts.

The Second Bid Round—December 11 and 12, 2009

After learning some important lessons from the first bid round, Iraq and the international oil companies were better prepared for the second round. A new venue was chosen for the second round—the oil ministry headquarters. Shahristani continued with his plan to conduct the second bid round at the oil ministry on December 11 and 12, even though there were several vehicular explosions throughout Baghdad early in December. A red carpet was installed from the front of the building through the hallways and to the large auditorium in a wing of the ministry headquarters. More than forty companies participated from twenty different countries. The oil company representatives sat on one side of the auditorium. There were seven American companies: ExxonMobil, Chevron, Conoco Phillips, Hess, Marathon, Anadarko, and Occidental Petroleum. Again, as with the first bid round, the dignitaries sat in front and included Prime Minister Maliki, Oil Minister Shahristani, Thamir Ghadhban, senior leaders of the oil ministry, and several ambassadors and diplomats from the embassies in Baghdad. They sat on the side of the auditorium opposite the oil companies. The diplomats were from France, Turkey, Russia, Japan, Malaysia, the United States, the United

Kingdom, and the Netherlands. The clear glass box was placed on a table on the stage. Seated close to the bid box were members of PCLD and their contractor, Gaffney, Cline & Associates.

The oil reserves being auctioned during the first and second bid rounds amounted to more than eighty-four billion barrels of proven reserves. The global energy sector had never experienced anything similar in the past and will probably never again in the future. Shahristani highlighted in the days before that an acceptable standard for remuneration fees had been set during the first bid round. The fields offered were more acceptable to many Iraqis inside the oil sector. These fields had little development work completed compared to the fields offered in the first bid round.

The first field for auction was the Majnoon oil field. This supergiant oil field is located just north of Basra with reserves of more than thirty-eight billion barrels. The winning bid went to the consortium of Shell and Petronas. Their bid was for a remuneration fee of $1.39 and a production plateau target of 1.8 million barrels per day.

The second auctioned oil field was the West Qurna phase two project. The West Qurna supergiant oil field has more than forty-three billion barrels of recoverable reserves. Phase two is located west of Basra on the northern edge of the West Qurna phase one project, which was auctioned to Exxon-Mobil in the first bid round. The winning bid went to the consortium of Lukoil and Statoil. Their bid was for a remuneration fee of $1.15 and a production plateau target of 1.8 million barrels per day.

Seven oil fields were successfully auctioned of the ten offered during the two days. In addition to the two fields just mentioned, the remaining five successful oil fields were Halfaya, Garraf, Badra, Qaiyarah, and Najmah. The oil companies were not interested in the other fields. Either the security was poor or the investment terms did not offer enough incentive for them. There were no bidders for the following three projects: East Baghdad, the Middle Furat oil fields (Kifi, West Kifi, and Merjan), and the Eastern Fields (Gilabat, Khashem Al-Ahmar, Nau Doman, and Qumar).

The second bid round went much smoother than the first round. Minister Shahristani was in a much better mood. He started joking with some of the bidders during the bid sessions. Most notable was the representative from Sonangol, who attempted to get a higher remuneration fee on the Najmah oil field accepted by Shahristani. He grabbed the microphone and said to the oil minister, "We expected a little bit higher. Can you go a little bit higher?" He was negotiating as if he was at a bazaar negotiating for rugs. He eventually accepted the $6.00 maximum fee as the people in the auditorium began to laugh. It was an amusing scene, a humorous relief from a very serious occasion that the global energy sector will probably never experience again. Iraq was auctioning one-third of the country's proven reserves over the two-day bid round.

Iraq had signed twelve contracts over the previous year with some of the largest oil companies in the world to develop their oil fields (see Table 18.1). These were all twenty-year technical service contracts that required the oil company to meet a plateau production target within a timeline of either six or seven years. Incremental production of 11.7 million barrels per day would be achieved if each of the twelve consortium operators successfully met their contract targets. Iraq still had other oil fields operated by the oil ministry producing almost two million barrels a day. These included Kirkuk, Bai Hassan, and several smaller fields. The contracts indicated that it was possible for Iraq to produce over thirteen million barrels per day by 2018, five times more than they produced in 2009. Saudi Arabia produced about ten million barrels per day in 2015.

The contracts may have stated such large production figures, but no one believed it possible to achieve those targets within seven years. First, the logistics for achieving such volumes would more than strain the port facilities and the rest of Iraq's infrastructure. Second, the oil ministry would resist the higher costs associated with the high demand for limited oil development resources in the Middle East, not overpay. Third, there was a high probability that some of the operators would not be able to meet their contractual commitments due to security issues. Fourth, the export infrastructure was identified as the biggest bottleneck for achieving the high volumes within the target time period. And finally, the market impact had to be considered. Thamir Ghadhban would highlight many times that it was Iraq's objective to optimize revenues, not production.

Oil Minister Shahristani had more than answered his critics about Iraq's declining production. His plan of executing technical service contracts with some of the world's largest oil companies silenced those criticisms.

The Third Bid Round—October 2010

The first two bid rounds did not achieve a contract for any of the gas fields. The first bid round included the Akkas and the Mansuriya gas fields, but no acceptable bids were received. The country would potentially produce more than enough associated gas from the oil fields, but dedicated gas production for supplying domestic power plants and the petrochemical industry was preferred. A third bid round took place in October 2010 to address this issue. Only smaller oil companies showed interest in the gas fields, so it did

Table 18.1 First Two Bid Rounds and Ahdab Oil Field

Field	*Consortium Operator*	*Duration to Reach Plateau Production Target (years)*	*PPT (barrels/day)*
Ahdab	CNPC	7	140,000
1st Bid Round			
Rumaila	BP (51%)	6	2,850,000
West Qurna (Ph 1)	ExxonMobil (80%)	6	2,325,000
Zubair	ENI (44%)	6	1,200,000
Maysan oil fields	CNOOC (85%)	6	450,000
2nd Bid Round			
Majnoon	Shell (60%)	7	1,800,000
West Qurna (Ph 2)	Lukoil (75%)	7	1,800,000
Halfaya	CNPC (50%)	7	535,000
Garraf	Petronas (60%)	7	230,000
Badra	Gazprom (40%)	7	170,000
Qaiyarah	Sonangol (100%)	7	120,000
Najmah	Sonangol (100%)	7	110,000
Total contracted plateau production target after seven years (2018)			11,730,000

Data Source: Iraq Oil Report

not have the fanfare of the previous two bid rounds. Fifteen companies had prequalified, but only five showed in Baghdad for the auction. Three fields were successfully auctioned: the Akkas, Siba, and Mansuriya gas fields.

The Akkas field is located in Anbar Province, near the Syrian border. The winning consortium included Kogas of Korea and KazMunaiGaz of Kazakhstan, with a remuneration fee of $5.50 per barrel oil equivalent produced (boep) per day. Their plateau production was four hundred million standard cubic feet per day.

The Siba gas field is located just south of Basra. The winning consortium included Kuwait Energy and TPAO, with a remuneration fee of $7.50 per boep per day. Their plateau production was one hundred million standard cubic feet per day.

The Mansuriya gas field is located in Diyala Province, northeast of Baghdad. The winning consortium included TPAO, Kuwait Energy, and Kogas, with a remuneration fee of $7.00 per boep per day. Their plateau production was 320 million standard cubic feet per day.[2]

The Fourth Bid Round—May 30 and 31, 2012

Exploration work was completed over less than 50 percent of Iraq. The fourth bid round took place in Baghdad on May 30 and 31, 2012. Unlike previous bid rounds for known oil and gas fields, the fourth round was strictly to auction twelve large blocks located throughout the country to explore for oil. Only three of the twelve blocks received acceptable bids. An industry consultant attending the auction stated, "Expectations were low going into this, as the terms were so fundamentally unsuitable, but I am slightly surprised by how bad the turnout was." The winning bidders included block 8 to Pakistan Petroleum Company, block 9 to a group of three small Middle East oil companies led by Kuwait Energy Company, and block 10 to a Lukoil/Inpex group.

The bid rounds had achieved their objectives. Politically, they saved the political career of Dr. Hussain Al Shahristani, and most probably Nouri Al Maliki. Without two successful bid rounds, Shahristani definitely would have been gone in the spring election of 2010. Maliki probably would have fallen too. However, not only were both rewarded with four more years, but Shahristani was promoted to the new position of deputy minister for energy, responsible for oil and electricity. This position had never existed; it was created as a reward for Shahristani. It was also an attempt to bring more focus to the Ministry of Electricity. His portfolio was the Ministry of Oil and the Ministry of Electricity. For the oil sector, the bid rounds achieved the quick return of the international oil industry to Iraq. Not just the oil companies but also the oil service companies returned to Iraq. It was the drillers, loggers, and other service companies that had refused to work directly for South Oil Company. Weatherford was the only service company in the south with drilling rigs before the bid rounds and its presence was limited. Schlumberger, Halliburton, Weatherford, and the like quickly mobilized drilling rigs to respond to the multibillion-dollar demands from the international oil companies.

Shahristani had developed an expertise within the oil ministry headquarters at auctioning oil and gas fields. He and PCLD did this very well. His next challenge would be to execute these contracts and manage the growth competently. Unfortunately, he did not have the team at the ministry to do this well and refused to bring in the right person that could.

The selection of a new oil minister fell to a committee of key members of parliament in the new government. Shahristani headed the committee. Several Iraqis advised me that the committee wanted Thamir Ghadhban to lead the oil ministry. However, Shahristani vetoed everyone and appointed his puppet, Abdul Karim Luaibi, in early 2011.

Four of Luaibi's former supervisors and peers assessed his career this way.[3] Luaibi was a senior engineer in the Technical Directorate prior to 2006 responsible for procuring chemicals and catalysts. Most of these procurements prior to 2003 were accomplished with nefarious foreign agents, brokers, traders, and suppliers under the Oil-for-Food program. He was not a stellar performer but always seemed to procure the chemicals needed by the refineries and production companies. They assessed his potential as an employee who would probably continue in that same position until his retirement from the oil ministry. In 2006, his relative became prime minister and suddenly his experience became highly valued by the new government. He was quickly promoted within the oil ministry to inspector general, upstream deputy minister, and then oil minister within a few years. I personally met with him several times while he was inspector general and deputy minister. His English-language skills were poor, so it was not possible for me to assess his skills. He seemed to be a likeable guy and always wore a very nice, expensive-looking tailor-made business suit. Could he handle the challenges of competently executing the bid round contracts? Let us look at the facts.

The international oil companies moved quickly to execute their contracts. One of their first contractual requirements was to identify the base production volumes for their specific oil field. This number was important because future profits would be determined according to this baseline production volume. BP signed the first contract, and so they were the first to figure out a way to determine the base number. They learned that South Oil Company had no operational meters and the storage tanks were too limited to provide a good way to calculate daily production. South Oil Company had been estimating the production of the Rumaila oil field.

BP and South Oil Company could only agree to baseline production numbers within about 10 percent of each other. BP's estimate was close to one million barrels per day, while South Oil Company's estimate was close to 1.1 million barrels per day. After a few attempts that arrived at about the same disagreement, they agreed to a compromise. The British executive added a few barrels to the compromise number to make it a memorable 1,066,000 barrels per day in acknowledgment of the Battle of Hastings and the Norman conquest of England in the year 1066.

Determining baseline production numbers for each contract may have been the easiest challenge for the incoming international oil companies. Their initial challenge was just getting visas for their employees to enter the country. Iraq was not ready to process hundreds of oil employee visas in 2010. It did not have the capacity at its ministries in Baghdad and was well understaffed at the Basra Airport immigration office for processing foreign workers. Delays for visas were several months long. Oil Minister Luaibi seemed ineffective in helping. Deputy prime minister for energy Shahristani seemed to show empathy for the oil companies, but he too was ineffective. Even with all the high-powered Iraqi government focus, this was still a significant problem for the oil companies when I departed Iraq at the end of 2011.

Once the employees received visas and entered the country, the security of those employees became a high priority. There was always a threat of kidnapping for the oil company employees in the Basra area. Fortunately, there were several private sector security firms that had been working throughout Iraq for the previous six years under contracts with the US government or other governments. So, the movement of employees to oil fields and other locations was not as big of an issue as finding secure office space and lodging for those employees. Again, BP took the lead.

The UK consulate in Basra was colocated on the US forces compound adjacent to the Basra Airport. BP was able to get office space and limited lodging from the UK consulate, which was a tenant on the US forces base. Shell would also get office space the same way. This provided secure offices and lodging as a starting point until BP and Shell could find or build more suitable space elsewhere in the south.

An amusing incident occurred when ExxonMobil approached the US embassy in Baghdad to request temporary office space in Basra on the US forces compound. The embassy initially declined ExxonMobil's request, saying that they could not show any commercial favoritism to an American private sector company. We let the embassy officials know that BP and Shell were already subtenants under the UK consulate on the same compound, so they found a way to accommodate the ExxonMobil request.

A significant technical problem was the availability of water for injection into the oil field reservoirs. Several of the international oil companies recognized in 2010 that they could never meet their contractual production plateau targets without getting access to additional supplies of water. This water was required for injection into the reservoirs thousands of feet below the surface to maintain the high pressures. Special injection wells were drilled and compressors installed on the surface to accomplish this process. About 1.5 barrels of water were needed for injection for every barrel of oil produced. Maintaining the high reservoir pressures enabled a high rate of crude oil flow to the surface through the production wells. Limited amounts of water were available from the Qarmat Ali water processing plant in Basra. This water was primarily earmarked for the Rumaila and Zubair oil fields. West Qurna, Halfaya, Majnoon, and several other oil fields would need access to significant volumes of water.

To address this issue, the oil ministry approached ExxonMobil in 2010 to lead a multibillion-dollar project called the Common Seawater Supply Project (CSSP). ExxonMobil designed a modular system that would enable each international oil company to construct their own pipelines and complete their specific hardware at a central facility close to the Umm Qasr cargo port. It was a good approach for having a common design and allowing each company to construct the hardware, as needed. The ExxonMobil project management team held several meetings with South Oil Company as well as other operating companies in the south to explain the design and the approach.

The Ministry of Oil complained of the high costs associated with the project and delayed approvals until it had studied and reviewed every detail of the specific costs. Meanwhile, ExxonMobil could not get acceptable cost recovery for even its limited CSSP planning work. The project was at an impasse; nothing was moving forward. It was stuck in the ministry.

During the summer of 2011, I met with the executives of BP and ExxonMobil assigned to Iraq at a compound inside the Green Zone. They were interested in my assessment of the timing of the SPM export project due to start up in late 2011 or early 2012. The operating companies were anticipating production cutbacks due to limited export capacity before the project's start-up. We also discussed several of the common problems still facing the operating companies. I tried to maintain communications with most of the international oil companies to help me better explain their issues to the US embassy and my contacts within the Iraqi government.

Delays in getting visas still existed. Although some improvements were made, additional progress was needed. There were administrative delays to project approvals by the oil ministry that were becoming of greater concern. This one did not surprise me because I knew that the oil ministry saw too much production coming onstream without enough export capacity. Its process for delaying this additional production would be to slow down approvals. The CSSP probably fell into this category. The oil companies were also learning that the ministry was denying reimbursement on some types of incurred costs. This would directly reduce their profits.

The most significant issue was cost recovery, payments to the oil companies. The oil companies had the option of getting paid in cash or crude oil. A few of them considered getting cash, but everyone eventually opted for payment in oil. The failure of the government to pay the oil companies in a timely manner was not being addressed. This payment issue was a significant concern because the original contracts assumed that cost recovery would be immediate. The assumptions that determined a 20 percent return on investment required immediate cost recovery. Delays in recovering costs would have a negative impact on the return on investment and net present value targets. All of these problems were becoming a source of frustration for the international oil companies. These executives from BP and ExxonMobil were sharing their frustrations. The ExxonMobil vice president Jim Adams looked at me after we summarized the issues and stated, "Gary, we never would have signed this West Qurna contract if we had known in 2009 what we know today."

Jim's comment surprised me. For an ExxonMobil executive to say this in front of outsiders, especially their competition, said a lot. This told me that he was sharing this message with many people and that the message must be the feelings of his leaders back in Dallas. It was a clear communication that ExxonMobil considered the status quo unacceptable. They were making changes.

A few months later, in October 2011, ExxonMobil signed a production-sharing agreement with KRG. This was shortly after my final departure from Iraq. The agreement with the Kurds irritated Deputy Prime Minister Shahristani and Prime Minister Maliki. The Kurds had signed several contracts with oil companies since 2004 and those oil companies were all blacklisted by the Baghdad government. They could not buy Iraqi oil or participate in the bid rounds. ExxonMobil's decision defied the intent of the Baghdad government to control all of Iraq's oil in Baghdad. Why did ExxonMobil take this strategy? How would Shahristani respond?

I saw an ExxonMobil vice president a few weeks later at the annual Army Navy football game. We were enjoying a great pregame tailgate party at FedEx Field in Maryland. Pat and Bob MacArevey of the West Point classes of 1978 and 1972 put the tailgate together. They were well known for their Army football tailgates and this was a beautiful, warm day with hundreds of people socializing. The vice president had some subordinates working in Basra at the time. I brought up the subject of the KRG deal and expressed some surprise about the bold move on the part of the executive committee. I received a good response: "Gary, you know full well that ExxonMobil loves to do business in countries where we are welcomed, are appreciated, operate within a legal framework, and are encouraged by the local government to earn a fair return on our investment. We feel welcomed and appreciated by the KRG government. At this point, we cannot say the same thing about the Baghdad government." Unfortunately, Army lost the football game 27–21 and Iraq was losing one of its best operating companies.

Shahristani removed ExxonMobil as the project lead on the CSSP. ExxonMobil announced its intent to sell its share in the West Qurna oil field contract. In December 2013, CNPC of China and Pertamina of Indonesia purchased more than half of the ExxonMobil share. ExxonMobil met several times with the Iraqi government and decided not to sell out totally. It kept part of its ownership share and decided to maintain its role as an operator of the field. I was personally happy to see them remain as the operator. The South Oil Company managers and engineers who worked with me on the export project described the working relationship between ExxonMobil and the Iraqis as the best among all the contracts in the south. The ExxonMobil employees did a nice job of explaining technical and business decisions, much better than the rest. A few of those ExxonMobil employees had been US military officers who had worked with me in Baghdad in 2003. They left military service after 2003, joined ExxonMobil, and eventually returned to Iraq as oilmen to help rebuild the country. They were playing an important role in the development of the West Qurna oil field.

Very little was accomplished on the CSSP in the four years following the removal of ExxonMobil as the project lead. Discussions were initiated between the oil ministry, Petro China, and ExxonMobil in 2015 to rejuvenate the project, but nothing definitive resulted. It is widely understood that production in the south is limited by the amount of water available for injection into the reservoirs and that the current production is approaching that ceiling.

After 2010, other significant changes were made to the original bid round contracts. Sonangol declared force majeure on its two contracts in the northern oil fields of Najmah and Qaiyarah to release the company from its contractual obligations. "Force majeure" is a contract term and refers to unforeseeable circumstances that prevent someone from fulfilling a contract. A force majeure clause was in all of the oil field development contracts. The security situation became too difficult and the Baghdad government released the company from its contract. The two northern gas field contracts at Akkas and Mansuriya with Kogas of Korea and TPAO of Turkey also declared force majeure due to security problems. There are no current plans to restart work on any of these three northern fields. Other bid round contracts were renegotiated to lower the target maximum production volume and extend the period of the contract. Similar to ExxonMobil's equity position in its West Qurna contract, some of the other oil companies bought and sold equity positions in their original bid round contracts.

The bid rounds demonstrated Shahristani's political genius. He survived the attacks of his enemies and not only survived but thrived. He learned how to go back to get the support of his political base in Najef during his darkest hours of 2008. Whether he intended it or not, he made his temporary failure in the first bid round a huge problem for the Maliki government, forcing emergency actions not normally taken by the Council of Ministers. He turned failure into success. Time will tell and the Iraqis will have to determine if Shahristani's short-term political success translates into a long-term win for Iraq. He negotiated away much-needed cash flow for his country in return for low, fixed remuneration fees that were politically popular. Such a strategy was excellent during periods of high oil prices. The low global oil prices of 2015 highlighted a huge shortcoming of his contracts. As oil prices fell, his fixed remuneration fee became a much higher portion of the country's revenue share. The contracts assigned much-needed cash to oil field development rather than to other demands on the government. This prompted a lot of criticism, even from his political base.

CHAPTER 19

Leaving Iraq

September 2011

BOOM!!! VP Biden must be visiting again.
—Kevin Ross

It was August 2011. My Etihad Airlines flight was landing at the Baghdad Airport for my final trip into Iraq. We pulled up to the commercial terminal and all the passengers exited. I was the only American on this flight from Abu Dhabi. Most passengers were Iraqis returning from trips outside the country. There were half a dozen non-Iraqis that looked like foreign oil company employees. The process through immigration was slow but much faster than just a few years earlier.

My security team met me on the other side of baggage claim. It was a private sector security group of former UK military folks along with their local Iraqi employees. The two vehicles were armored SUVs with two armed bodyguards and a few weapons inside the vehicle. They put a protective vest on me before getting into the back seat of the vehicle. The US military was leaving the country and most DOD contractors had already departed. This last contract that Kevin Ross and I signed for 2011 required that we get our own ground transportation to and from the airport. The cost for such secure taxi service was $2,200 one way. It was the most costly taxi service I've ever used. Our security company was not the most expensive, but it was not the cheapest either.

The trip from the airport to the Green Zone was similar to the others. The Doura power plant had only two of the four smokestacks operational. This prompted my recollection of my first trip from the airport when only one smokestack was operational. Back then, I wondered how many weeks it would take to get all four smokestacks operational. Over the previous eight years, I never saw all four in operation, I seldom saw three, and most of the time there were only two smokestacks operational. Our reconstruction effort seemed to take a similar pattern, two steps forward followed by one step backward. My thoughts moved to a comparison of my first tour in 2003 versus my second tour, which was coming to an end.

The pace of my first tour was a dead sprint. We worked sixteen hours a day, seven days a week, with a one-week break every three months. That was a draining experience that had me wound very tight when I departed in July 2004. Such a pace could not be maintained during my second tour. This tour lasted almost five years, and I was able to avoid the pressures of my first tour. As an independent contractor on this tour, I did a much better job of managing my time. First, my time outside Iraq was more frequent and longer. The other significant difference was that my evenings were considered my escape time. My Amazon book library in Baghdad grew. Each night I tried to get to the gym for a workout while watching Stephen Colbert on Comedy Central. After my workout, I returned to my room to watch an hour of Jim Cramer and *Mad Money* that I downloaded through iTunes. These gentlemen provided me with a wonderful daily mental escape from the rigors of a war zone.

During the ride to the Green Zone, my thoughts shifted to the US government oil organizations over the last few years and the great Americans in those groups. After the Energy Fusion Cell, we morphed into the Energy Services Division under the Multi-National Force–Iraq. We physically moved from the Freedom Towers to the new US embassy compound located in a different area of the Green Zone. Rear Admiral Mark Belton led the new division with Colonel Rick Waddell as the oil lead. Rick was replaced by Commander Bobby Quintos, and then by Air Force Major Vern Starman, followed by Major Jeff Steinbrink and Major Chris Dube. Air Force Brigadier General Bob Ranck replaced Admiral Belton, and Colonel Greg Zehner replaced him. The Multi-National Force transitioned to US Forces–Iraq at the beginning of 2010, after the UK forces departed Iraq in December 2009. General Odierno lost his UK deputy commander, so US Lieutenant General Ken Hunzeker deployed to fill that job.

The oil group was always under the oversight of the deputy commander. Now that we had a US deputy, nothing changed. Hunzeker requested a briefing on the ICOEE project shortly after his arrival. He had just completed his previous Iraq tour a few years earlier as the commanding general, Civilian Police Assistance Transition Team, and advisor to the Minister of Interior. It was an informal brief in his office. I took a deck of PowerPoint slides to review the current status of the export project. Hunzeker had a financial background, having worked several assignments in budgeting at the Pentagon. After going through the first PowerPoint slide, he asked three questions that were answered. As I turned to the second slide, he said, "Gary, I got it. You don't need to cover anything else. That project is the financial life to this country. I cannot think of a more important project. What can I do to help you?"

It was good to work with a general officer that was a quick study. Ken would leave the army after this tour and join the corporate world. He quickly moved into the CEO position of a publicly traded company. It was a pleasure to have such a talented senior leader advocating the export project at a crucial time. His replacement, Major General Ed Cardon, was also a quick study. He was a Corps of Engineers general officer that would move on to take over the army's Cyber Command several years later. These talented senior military leaders that were on their second or third tours in Iraq seemed to know how to add much more value to the Iraq mission than their counterparts on their first tour. This was also true of the senior State Department leaders assigned to the embassy.

My expensive taxi was now approaching the entrance to the Green Zone. The security for the Green Zone had transitioned recently to only Iraqi military. The US forces had moved off of the checkpoints. The United States had started demobilizing the year before when General Lloyd Austin took command of US forces from General Ray Odierno in October 2010. Austin's mission was to demobilize and safely remove the US military and equipment from Iraq before the end of 2011. He was on a glide path to accomplish that mission before Christmas. Kevin Ross and I were the last two DOD independent contractors. Our contracts ended September 30, 2011.

We pulled up in front of the embassy compound and I got out. The compound had a fifteen-foot wall with four or five entrance gates on the three-kilometer perimeter. The construction was completed in May 2008 and it opened for business in January 2009 at an estimated cost of over $750 million. There were a few thousand Americans living and working on the compound at the time.

It was a hot summer day, as all August days are in Baghdad. I quickly processed through security at the gate and walked inside the walled compound. As I proceeded toward my current accommodations at the DOD trailer park, I phoned Kevin Ross to let him know that his relief had arrived. He agreed to meet me at our trailer and join me for lunch. It was a huge compound with four or five large apartment buildings that housed the State Department employees and their contractors. Other buildings included the main Chancery, office building Annex I, office building Annex II, and the recreation center. The Chancery and Annex II had additional security at the entrance so that only personnel with the proper security clearance could enter those buildings. Classified work areas existed in both buildings. Annex I contained the counselor services for processing visas for Iraqis wanting to visit the United States. The permanent buildings were all constructed with blast-proof windows and reinforced construction designed to withstand mortar and small-rocket attacks. They proved their efficacy more than once during my time.

All of the military personnel and DOD contractors lived in the DOD trailer park adjacent to the dining facility at one end of the compound. Kevin and I shared one of the two-man trailers, also known as a containerized housing unit (or CHU) to the military. They were small, functional lodging units with a minirefrigerator and television. A reinforced steel covering was constructed over the trailer park to protect us from mortars and rockets. Fortunately, no rockets ever landed on that covering to test the system.

Kevin was glad to see me. He would be leaving the next morning for his final trip out of the country. I was there to relieve him for the last time. We spent the day going over those items that we could not discuss over the telephone while I was out of the country.

Kevin and I had worked in the Annex II building when we first transferred to the new embassy compound a couple of years earlier. There was still a sizeable military presence and mission then. We moved into the Chancery in late 2010 to support the embassy Econ section when many of our military teammates demobilized. Our contract remained with DOD, but we took daily direction from the embassy.

The US government's interest in Iraq's oil sector had significantly changed. It was now limited to observing and reporting back to the State Department. Kevin and I provided expertise for the Foreign Service officers writing those reports and trained them on the important points of the oil sector. A new embassy staff arrived every summer and our work would soon be complete after we trained the new group that had just arrived. Our work was finished. It was time to leave.

"Allahu Akbar. Allahu Akbar."

I woke up to the call to prayer the next morning at just before 0500 hours. There was a mosque with excellent loudspeakers that could be heard all over our section of the Green Zone. My brain would register that it was almost time to wake up. I had another sixty minutes of sleep as I rolled over in bed.

BOOM! A rocket just exploded nearby. It sounded like it was inside the embassy compound. My extra hour of sleep was cut short.

Kevin woke in his bed and commented, "VP Biden must be visiting again."

We laughed as we wondered how the Iraqi rocket man always seemed to know when our vice president was visiting the embassy. His visits were always kept confidential for security reasons. We were never told in advance, but it seemed like the rocket man always got informed. Kevin said that it was the first rocket attack in more than a month. I hoped that it would be the last before my departure.

Vice President Biden had Iraq as part of his portfolio at the White House. He visited Baghdad quite often. Secretary Clinton visited Iraq in April 2009, just after becoming secretary of state. I do not recall her returning.

Kevin and I went to breakfast together and he said his final farewell to many that we saw at the dining facility. A more formal farewell party had been held for him the previous evening at the Baghdaddys Club on the embassy compound. Baghdaddys was the watering hole that stocked many popular adult beverages. It was the embassy's social club that opened most nights with volunteers as bartenders.

After breakfast, I helped Kevin with his bags and walked him to the gate, where our secure taxi service loaded his luggage and took him to the airport for his final departure. Kevin spent twelve months in our CPA oil section during 2003 and 2004, when he was mobilized as a navy commander in the reserves. As a submarine officer with a nuclear engineering degree from Texas A&M, Kevin was a very quick study of the oil ministry that year. He returned to Iraq as a contractor in May 2008 when we needed help with the Energy Fusion Cell. The two of us had been working back-to-back tours over the last couple of years. He would be missed.

The embassy chancery was a remarkable building, standing four stories high. It was the first building after entering the main gate of the compound. Many of the embassy offices were on the second floor. The US military had their offices on the third floor in 2011. The fourth floor had other agencies and support staff. The Econ section was located just down the hall from the ambassador's office on the second floor. I entered the office and went to Jim Turner's cube. Jim had just arrived in country earlier in the summer with the latest team of Foreign Service officers. He was an experienced senior Foreign Service officer, heading up the team that monitored the oil, electricity, and water sectors. Although he had worked on the Near East Asia desk at Foggy Bottom, this was his first tour in Iraq.

The State Department sent many excellent Foreign Service officers over the last few years for a twelve-month tour. As the oil sector subject-matter expert, I was fortunate to work with many of these great Americans. Some of those names were Jeff Izzo, Ken Fairfax, Mike Uyehara, Mike Dodman, Ed Parks, Karen Choe-Fichte, Patrick Dunn, Eric Camus, Dale Eppler, Dianna Chianis, Erik Ryan, Larry Milam, Kit Redmer, Brian McFeeter, Bob Murphy, Tim Phillips, Seneca Johnson, and Nicholas Papp. The State Department also hired subject-matter experts for electricity and water that had been in the country since 2005. Those guys had wonderful experience in the private sector and were great to work with. The electricity experts were John Dempsey and David Ensign. John Thompson was the water expert.

Entering Jim Turner's cube and greeting him, I asked, "Jim, when did the vice president arrive?"

Jim whispered that the vice president's visit was very confidential.

I smiled and told him the theory that had developed over the last couple of years about the relationship between visits from Vice President Biden and the early-morning rocket attacks. It was not a 100 percent correlation, but it seemed like it was very close.

We had a busy calendar planned for my last month in Iraq. The highlights included final trips to the north and to the south. The planning would be a joint exercise between the Econ section and Major General Jeff Buchanan's team in the J-9 staff section. We were given a high priority. General Buchanan's senior civilian assistant, Greg Garcia, made the trip north with us.

Our trip north was a one-day Blackhawk flight. We flew from the helipad next to the embassy compound to visit the Bayji Refinery, which was followed by a helicopter flyover of the Mosul Dam. The refinery had made progress repairing the damage from the spring attack that had shut down several processing units, but the leadership was still shaken over that event. It was good to visit them, but my message was that the US forces would all be gone within a couple of months. It was a reiteration of what they already knew. That message was not good news for these Sunni technocrats. They had experienced success in 2007 and 2008 with the US Army providing the needed security. They were now on their own and they recognized their vulnerability.

Located on the Tigris River just north of Mosul, the Mosul Dam was the largest dam in Iraq. Its hydroelectric plant could supply 750 megawatts of power. This was enough electricity to power the city of Mosul. Construction of the Mosul Dam was an engineering challenge when constructed in the early 1980s by a European consortium. The soluble gypsum foundation required constant maintenance. The Iraqis had to inject grout into the cavities formed when the gypsum dissolved. A scare about a possible dam collapse occurred every time the grouting program was interrupted due to wars or insecurity. The chance of a catastrophic collapse was unknown. The engineering community differed on such an event, but they all agreed that a grouting program was essential for the long-term health of the dam. The water levels were intentionally kept low over the last couple of years because of the threat of failure. The lower levels impacted the amount of power generated by the hydroelectric plant.

For our trip south, we flew out of the VIP terminal at the Baghdad Airport on a C-130 US air force aircraft. The VIP terminal was the site of my first arrival in Baghdad in 2003. Things had improved significantly at this terminal over the last eight years. It was still operated by the US air force, but it now had a nice lounge with a refrigerator, a television, magazines, and fully functional restrooms.

We linked up with two Blackhawk helicopters at the Basra Airport. They flew us south over the waters of the northern Arabian Gulf to view the construction activity for the offshore export project. We got some great pictures of the large pipe-laying barge as it was welding pipe joints and floating the pipe toward shore over the eight-kilometer dredged canal. We could not fly over the Al Fao terminal because it was too close to the Iranian border. The pilot said that a five-mile restriction had been in place for more than a year as part of the demobilization protocol.

From there, we flew to the Umm Qasr cargo port and along the Khor Al Zubair small oil port facilities. These ports were used to import gasoline, diesel, and LPG products. We flew over the Zubair and Tuba pump stations, and then to the lay-down yards of drilling companies full of unused rigs. There were only two drilling rigs in the entire south just three years prior. During this flight, we saw three rigs in one storage yard. We flew west to the Rumaila oil field and flew from south Rumaila north to the West Qurna oil field. There were several operational drilling rigs as well as gas-oil separation plants, and many flares burning associated gas from the oil production.

Iraq still flared most of their associated gas that was produced with the oil. It would take much more equipment and billions of dollars to capture and transport these gases to power plants or petrochemical plants.

The advancements of the export project, rebuilding pump stations, drill rigs, and much more economic activity were visible compared to the time of my flyover in 2009. Any oilman would be excited with what we saw during this flyover, especially knowing the history of the last eight years. My feelings were quite different after touring the southern infrastructure compared to the northern infrastructure. I was sad after visiting Bayji, but I had a feeling of joy and satisfaction after the southern visit. The fruits of our efforts on the export project in the south should help Iraq for decades to come. I had the opposite feeling after the Bayji visit.

Saying farewell to my Iraqi friends became physically difficult. The security situation dictated very conservative guidance from the embassy security office. I could not get out of the Green Zone and my Iraqi friends could not get to me. I visited Thamir Ghadhban a couple of times before my final departure. He was working and living inside the Green Zone at this point and invited me over to his home.

The embassy Econ group put on a nice farewell party for me. Seneca Johnson was Ambassador Jim Jeffrey's chief of staff. It seemed appropriate to have someone from my original Iraq team, the Energy Infrastructure Planning Group (EIPG), present to bid me farewell. She and I had started this voyage together in November 2002 at the Pentagon. She was the State Department representative on the EIPG cell. Many Americans across different organizations had been directly involved in the Iraqi oil sector since 2002 (see Figure 19.1). Seneca had wanted to volunteer for the original ORHA team but was pregnant at the time. Her son was eight years old now, so she volunteered to work for Ambassador Jeffrey at this point. In my farewell talk, I reminded Seneca of her comment to me in 2002: "The road to hell is paved with good intentions." Major General Jeff Buchanan also joined the farewell. They gave me a very nice send-off.

My last evening was spent with a West Point classmate, Phil Lynch. Like me, Phil had spent multiple tours and many months in Iraq. He also spent a tour in Afghanistan. Phil was a lawyer assigned to the Department of Justice (DOJ) offices in Seattle, Washington. He was a retired army JAG officer, having retired in the early 1990s. His family was a military family. His wife, Paula, was an army reserve medical officer who was mobilized and spent a tour in 2003 assigned to a military hospital in Kuwait.

Phil played a major role in the trial of Saddam Hussein. During his first Iraq tour, he advised the Iraqi prosecutor who prosecuted Saddam. This was one of the most memorable trials in history, covered daily in the international press.

DOJ sent Phil back to Iraq during the surge to lead the Rule of Law program. Senior leaders in the State Department recognized his great work with this program. State requested that the Justice Department send Phil back to Iraq. He served another tour in 2011 in Iraq and a tour in Afghanistan in 2015, leading the Rule of Law programs in both countries. Phil epitomized the West Point motto of "Duty, Honor, Country."

My last trip from the Baghdad Airport at the end of September was full of mixed emotions. My plane took off at noon, and as I looked out the window at Baghdad below, my feelings varied. There was joy over the sense of the accomplishments of the oil sector. There were joy and pride for being part of the surge under Generals Petraeus and Odierno. However, there was a lot of sadness about the current security situation and the knowledge that there was a high risk that the gains of the surge would disappear quickly with the current sectarian policies of the Maliki government. Those policies were evident at the oil ministry. The ministry had been significantly politicized since 2006. It was not going to be reversed any time soon. The professional oil ministry that we experienced in 2003 was now history.

During my departure from Iraq after my first tour in 2004, my emotions were much different. On that trip, the oil ministry was in good hands. A professional oilman, Thamir Ghadhban, had been appointed as the oil minister. Ghadhban was surrounded by career professionals who had been promoted based on performance and competency. More progress on reconstruction

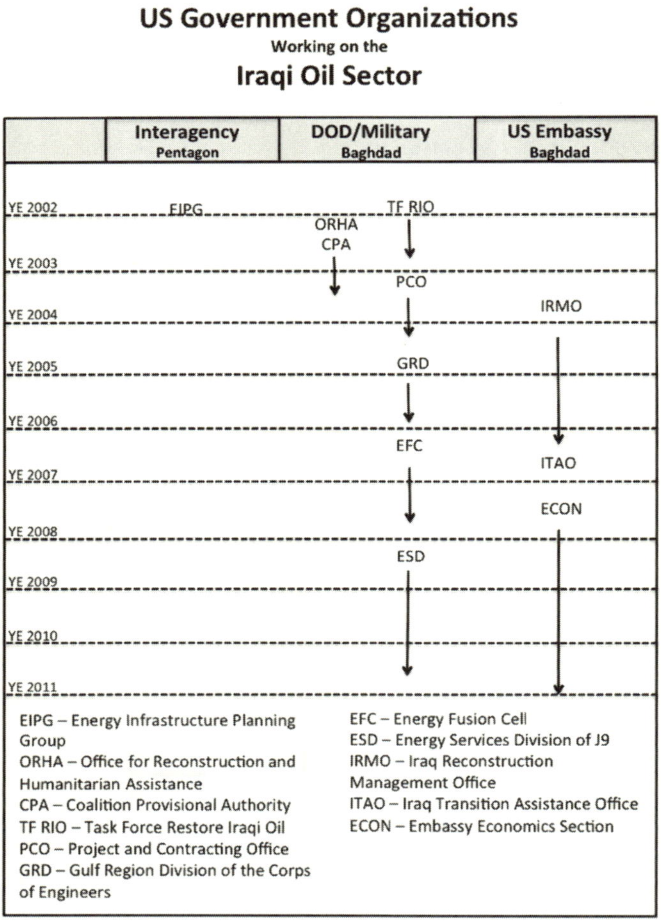

Fig. 19.1 US Government Organizations Working on the Iraqi Oil Sector

was needed, but such work would take time, and competent leadership would make it happen. I did not have those same feelings on this trip. The recent return of the international oil companies tempered some of my guilt. They would provide the backstop to help the oil ministry and Iraq when the oil ministry stumbled. That was my hope.

Four US service members made the ultimate sacrifice in September 2011 and six more were killed before Christmas. The total number would be 4,489 military casualties for the nine years, a much higher price than anyone envisioned in 2003.

CHAPTER 20

What Haunts Me

4,489 sons and daughters killed, 32,223 troops injured.

The costs to the United States for invading Iraq were huge. Some of those costs included 4,489 sons and daughters killed, 32,223 troops injured (not including PTSD), at least 134,000 Iraqi civilians killed, 150 reporters killed, 2.8 million people who either remain internally displaced or have fled Iraq, almost $1.7 trillion from the US Treasury through fiscal year 2013, and an estimated $490 billion of benefits owed to war veterans.[1] We achieved little to no benefit for these costs.

Could we have been misled into believing that WMD was the noble reason for this war? Or was it just an excuse? Could the war have been partly undertaken for the wrong reasons? Could oil have been one of those wrong reasons? There was a hidden oil agenda, but certainly not the one that most Americans imagined. What haunts me is that I may have been an unknowing participant in a secret oil agenda.

Our nation's second president, John Adams, was quoted as saying, "Facts are stubborn things; and whatever may be our wishes, our inclinations, or the dictates of our passions, they cannot alter the state of facts and evidence." I knew several facts about unofficial oil agendas in 2003, but they did not seem important at the time. They seemed to be just distractions to our mission. Many more facts have come to my attention over the last couple of years. My opinion has changed. These new facts are the source of what troubles me. The biggest is the linkage between two unscrupulous "dealmakers" and senior Bush Administration decision-makers. Specifically, Dr. Ahmed Chalabi was closely linked to the Bush Administration neoconservatives. That is well known, but his promise to deliver an oil deal is not. One senior neocon was linked to the notorious "King of Oil," Marc Rich. Rich made billions of dollars trading in nontransparent Middle East oil deals. By the 1990s he had been a symbol of unscrupulous oil deals for more than twenty years.[2]

The Chalabi oil deal was never consummated. Dr. Ahmed Chalabi's prewar promises amounted to nothing more than a duping of those senior Bush Administration decision-makers who trusted him. Chalabi deceived the neoconservatives about an oil deal and they sold the Iraq War to America. Chalabi's oil promise played a role in the decision for war. Was it a significant role? This is what plagues me.

It is human nature to remember the good things, the great experiences of life, and to disregard or file away the negatives, the bad experiences. As I looked back over my time at the Pentagon and in Iraq supporting Operation Iraqi Freedom, I wanted to look at the positives. I enjoyed the wonderful times that I shared with great American patriots and newfound Iraqi friends.

The interagency planning at the Pentagon brought together some of the best and brightest within our government. They were civilians and military; most were still active, but many had recently retired from government positions. Several of those individuals were from the Washington, DC, area, but there were also many brought in from outside DC. There were some that were from the Department of State, Department of Defense, Department of Energy, Treasury, Department of Justice, and the CIA. They were a great group of committed Americans, and I was privileged to have served with every one of them.

The movement into Kuwait and then into Iraq saw many new faces, but it matched the quality of Americans that I worked with at the Pentagon. My interfaces in theater with active-duty military increased. I worked with old friends on active duty; most were in senior leadership positions. My new friends were the young, bright, and energetic officers, noncommissioned officers, and soldiers that were the best that America had to offer. All of them were unselfish and dedicated Americans. No one bothered to ask them if we should invade Iraq, but they were there doing their job.

Phil Carroll's arrival in Baghdad in May 2003 was a personal highlight for me. Phil was an American oil industry statesman, a former senior executive, smart, articulate in his way, and, above all, a patriotic American. Other great oil industry leaders and patriots followed: Rob McKee, Mike Stinson, and Norm Szydlowski. The United Kingdom added two retired oil executives: Terry Adams and Bob Morgan. Australia sent a government energy expert, John Kjar. We put together my version of the "oil dream team" with these gentlemen and the US government civilians and military. Clarke Turner, Brigadier General Bob Crear, Frosty, Commander Kevin Ross, Colonel Tom O'Donnell, Major Mike Buck, Colonel Lem Dubose, Lieutenant Colonel Tony Cusimano, Robert Maguire, and several others were part of that team of 2003 and 2004. The list of senior military officers is endless: Petraeus, Odierno, Dempsey, Hunzeker, Helmick, and Strock were the most recognized names, but there were many others who did wonderful work without the equivalent recognition.

The list of patriotic Iraqis at the oil ministry with great oil industry experience is also endless. These gentlemen worked closely with Americans. They were Iraqis motivated to improve their own country. That list includes Thamir Ghadhban, Dathar Al Khashab, Jabbar Al Luaibi, Kefah Numan,

Mohammed Al Jibouri, Nabiel Lamosa, Dheyaa Al Bakka, Hazim Sultan, Dr. Radhwan Al Saadi, Dr. Faleh Al Khayat, Thair Ibrahim, Saleh Aziz, Sameer Michael, Hussain Al Fatal, Abdulilah Al Amir, Natiq Al Bayati, Riyadh Ghassab, Dr. Ahmed Al Barifkani, Kathan Al Anbaki, Faleh Al Kawaja, Akram Akrawi, Abdul Jabbar Al Waggaa, and Ahmed Al Shamma. These gentlemen were all patriotic Iraqis and competent oilmen that enriched my life during my first tour of 2003 and 2004. Others that became friends during my second tour included Dr. Ali Obaidi, Asri Moussa, Zaid Sandook, Dheyaa Al Jaffar, Fayadh Nema, Motasam Hassan, Nihad Moosa, and Oday Al Quoraishi.

The list of non-Iraqi professionals who worked with me between 2006 and 2011 is extensive. This list includes the great professionals at Foster Wheeler, Stanley Baker Hill, and Michael Baker Engineering.

The guidance of Robert Maguire before my second tour proved priceless. He recommended that I work on a project that was needed by the Iraqis, but that was off the radar screen of the embassy. His guidance was to avoid the politics inside the embassy. I threw myself into the southern export project. This led to the completion of the highly successful export project in the south, the ICOEE Project. My pride in contributing to that project is probably only surpassed by the tremendous pride shared with many great Americans who participated in the surge strategy of 2007 under Generals Petraeus and Odierno. Those successes provided new life to the Iraqi people and an opportunity to better their country.

The oil started to flow on April 23, 2003, and has never stopped, in spite of significant obstacles. Those obstacles were identified throughout the book and included US obstacles, such as political antagonism in Washington, a lack of military resources on the ground, political sensitivities to using Vice President Cheney's former company, CPA's payroll and budgeting policies, government procurement delays, optics around America's role in Iraq's oil sector, the Pentagon's Syria agendas, and others. Even the Iraqi obstacles of the insurgency, crime, and political decisions were successfully overcome to keep the oil flowing. The teamwork and partnering between the Iraqis and the coalition produced many successes. We overcame all the obstacles to keep the oil flowing. All of these positive thoughts dominated my thinking for a couple of years after returning from Iraq in 2011. Although the negatives were there on the periphery, I did my best to avoid them. My focus was on the positives and the progress of Iraq's oil sector.

That changed in 2013 when Ambassador Bremer convinced me to join him and Meghan O'Sullivan on a Rachel Maddow documentary. At the time, Meghan was the Jeane Kirkpatrick Professor of the Practice of International

Affairs and director of the Geopolitics of Energy Project at Harvard University's Kennedy School. After her CPA tour in Baghdad, she was appointed as a special assistant to President George W. Bush and the deputy national security advisor for Iraq and Afghanistan. Maddow's producers were putting together a documentary on the role of oil and the decision to go to war. Knowing that they would ask many questions about possible oil agendas, my preparation work addressed those issues. The result of the documentary and my research was my decision to start writing this book. Researching and writing the manuscript created many concerns about my experiences and my role.

The battle update assessments (BUAs) conducted by General Petraeus and his successors were previously mentioned. The surge brought hundreds of us together every morning to participate in, listen to, or watch the daily BUAs. They were every day at 0800 hours and broadcast throughout Iraq. It was easy to lose track of the day of the week in Iraq, but we always knew when it was Saturday. The Saturday BUA ended with the chaplain giving a memorial to one or a few of the fallen military heroes. The memorial normally consisted of a picture of the fallen soldier(s) with their name, their age, their hometown, and some other significant fact. It was a solemn part of the week.

Many times the age of the fallen would be the same age as my children—mid-twenties. The chaplain would talk about the soldier's family, their high school, sports, comments by fellow soldiers, and anything specific about this great American hero. It seemed like most were between eighteen and twenty-five years of age. There were many that were older with a wife and young children at home. A few had large families with many children and even grandchildren. My heart would always become very heavy during those memorials. A gigantic lump would suddenly appear in my throat that would seem to cut off my oxygen. My nose and eyes would water as if I was fighting a severe cold.

My thoughts would always shift from the fallen soldier to his family. How are they doing? How does a parent handle the death of their child? The worst fear of any parent is to have one of their children precede them in death. The parents of that twenty-five-year-old were probably the same age as my wife and me. They were experiencing a tremendous loss, the loss of a child—a child that they brought into the world, experienced in those early years, then in the teenage years, and then as a young adult. They were reliving how they had helped their child transition through life, school, playing football and baseball, and participating in activities such as scouts, fishing, hunting, and religious services.

How does a wife handle the death of her husband and the father of her young children? Captain Brian Freeman of Temecula, California, was one hero I will never forget. He was a 1999 West Point graduate, the father of two small children, and a Civil Affairs reserve officer that had been mobilized during the summer of 2006 and sent to Karbala, Iraq. He and his military team worked with the local Iraqis. Brian and a few of his colleagues were captured and later murdered by a group of pro-Iranian Shia insurgents in January 2007. My son-in-law did two military tours in Iraq; the second was in 2010. He and my daughter were so moved by the details of Captain Freeman's story that they named my oldest grandchild, Brian, after Captain Brian S. Freeman.

President Dwight D. Eisenhower's quotation about war would ring in my ears as the chaplain gave his weekly memorial: "I hate war as only a soldier who has lived it can, only as one who has seen its brutality, its futility, its stupidity."[3]

The memorials always left me very sad. My emotions got the best of me. A few times, my emotions turned from sadness to anger, especially during those early months of the surge when we were losing more than one hundred per month. I was angry over the original decision to go to war. I would remember Ambassador Robin Raphel's comment in Baghdad during the summer of 2003: "So there is no WMD, why are we here?"

I would recall the *New York Times* columnist Thomas Friedman's quotation. Friedman was a Pulitzer Prize–winning columnist who was highly regarded by our senior general officers in Iraq. His ability to ascertain information from all sides in Middle East disputes made his analyses much more credible than those of his peers. *Haaretz* is an Israeli newspaper with offices in Washington, DC. They interviewed Friedman in his Washington office in early April 2003. The war was only a few weeks old, but Friedman understood then what it would take many of us several years to fully comprehend: "It's the war the neoconservatives wanted. It's the war the neoconservatives marketed. Those people had an idea to sell when September 11 came, and they sold it. Oh boy, did they sell it. I could give you the names of twenty-five people who, if you had exiled them to a deserted island a year and a half ago, the Iraq war would not have happened."[4]

Friedman did not name the neoconservatives, but most were included in the hundreds of articles I have read on them since 2003. Those in government included Paul Wolfowitz, Doug Feith, Scooter Libby, John Bolton, Richard Perle, Elliot Abrams, Ken Adelman, Reuel Gerecht, Michael Ledeen, David Wurmser, David Frum, and others.[5] Those in the press included Bill Kristol, Robert Kagan, Judith Miller, and Laurie Mylroie.[6] One of them worked on my prewar planning team at the Pentagon, Mike Makovsky. Mike was the youngest person on our team, single, and the only team member who refused to go to Iraq. Only a couple of low-level neoconservatives took jobs in Iraq and they did not stay long, only two or three months. They left when things got tough.

Many of my colleagues familiar with the Iraq decision agreed with Friedman: if the neoconservatives had been shipped to a deserted island in 2001, we never would have gone into Iraq. How did such a limited few sell our nation of over three hundred million into fighting a war that would cost us so much?

My anger would be directed at the people in the Pentagon who cheered in Doug Feith's office when Secretary Powell made his key point about weapons of mass destruction to the United Nations. Why were Feith and his colleagues so happy about going to war? My anger would be directed at Vice President Cheney for what seemed to be his simplistic approach to Middle East politics—using our military to solve every Middle East problem.

My anger would be directed at President Bush for failing to require more certain information from the CIA before deciding on war. "It's a slam dunk" should have been immediately followed up with ten or more specific questions.[7] Such trivial comments without facts to back them up were irresponsible. My anger would be directed at the many congressmen and senators who gave their support to President Bush without asking those important questions that any competent lawmaker should ask before taking us to war. The common denominator of all the recipients of my anger was that not a single one of them was in Iraq with us at the time. None of their children or grandchildren was with us in Iraq at the time. None of their children or grandchildren was among the 4,489 fallen heroes of the Iraq War.

What was the role of oil in the decision to go to war? Many Americans since 2002 have asked that question, and the answers have surprised me. Many would comment that it was obvious, as if there was no other reason to go to war against Iraq, except for oil. Many would comment that Vice President Cheney pushed for war out of pure greed because of his former company, Halliburton. Some would comment that President Bush wanted to reward his friends in the oil patch with huge contracts in Iraq. I tended to ignore all of these comments as being politically motivated by the uninformed. I did not see anything close to that activity during my prewar planning and my time in Iraq. Phil Carroll commented to me when he arrived in Baghdad that if he saw anything close to an "oil grab" he would immediately depart Iraq. Phil stayed until the end of his tour in September 2003. Ambassador Bremer shared with me in 2016 that no one discussed oil before his departure to Baghdad in May 2003. It was discussed only in regard to its important role in Iraq's economy and the importance of getting the economy moving. Oil Minister Thamir Ghadhban responded to me in 2013 about such uninformed comments: “Gary, if the Americans were interested in taking our oil, you did a horrible job. The Chinese and the Russians have most of the large contracts.”

Some Americans would comment that the war was about oil because Saddam could use the money from oil sales outside of the UN Oil-for-Food program to pursue his weapons objectives. That comment made some sense. It was obviously true. But then, that supported the argument of weapons of mass destruction being the reason for war.

Short of reading President Bush’s mind, there is no way of determining if his reason for invading Iraq was motivated by oil. After the Maddow documentary, I took a different approach to thinking about the role of oil in the war decision. Rather than looking at accusations and dismissing them out of an obvious lack of factual evidence, I decided to look at all oil agendas that I saw during 2002 and 2003. I researched using my notes, my emails, Internet articles, and the dozen interviews that I conducted in 2015 and 2016. Most oil agendas that I found in my notes were quickly dismissed because there was no follow-up, no support for them. For example, a couple of inquiries came into Baghdad during the summer of 2003 from small oil companies and Republican supporters asking for oil contracts. Those companies were advised of the procedures that CPA and the oil ministry followed to award contracts. The individuals and companies were irritated with our response, but we never heard from them again.

Supplying Iraqi oil to Israel was the one unofficial agenda item that received a lot of government sponsorship and emphasis in the press. There was much more unofficial interest in this than in any other agenda item. Oil Minister Thamir Ghadhban stated in a presentation at a think tank in Washington, DC, during December 2005 that he was asked more questions about the Kirkuk-to-Haifa pipeline than any other oil sector issue. I did not become aware of many of the facts about this agenda until many years after 2003. I learned several specifics only during my manuscript research effort in 2015. Before 2014, I would have said that oil played little to no role in the decision to go to war. The more I learned of the facts behind the Israel oil agenda, the more I was disturbed and thought that oil played a significant role.

Dr. Ahmed Chalabi tried to convince the CIA and the State Department to buy in to his agenda for removing Saddam after 1991. Those agencies tried working with him but recognized that he was not trustworthy and eventually distanced themselves. Dr. Ahmed Chalabi had an unscrupulous reputation among his critics. His supporters would not defend him against this reputation. Chalabi found a more hospitable group in the powerful American conservatives with excellent contacts within the Israel lobby. These gentlemen were all very intelligent and all had experience in previous Republican administrations and in the Middle East. Chalabi failed with the CIA and State, but he had no problem convincing this group to support him. They were what I have previously called the neoconservatives, or neocons.[8] Why was it so easy for him to deceive the neocons into supporting his Iraq War agenda? What was he selling to them that he did not sell to the CIA and State? Was greed part of the equation? Milton Friedman had a theory that "societies run on greed." Let's try to find the possible greed involved in this oil agenda. Let's follow the money!

Let me lay out the facts that have contributed to my hauntings. These facts are grouped in two categories: those that I knew in 2003 and those facts I learned later.

Known by me during the summer of 2003:

- Senior political appointees at the Pentagon pushed for Dr. Ahmed Chalabi to be prime minister.
- The only deliberate infrastructure target during military operations was the K-3 pump station that pumped oil to Syria.
- Mike Makovsky at the Pentagon contacted me during the summer of 2003 trying to get a status on the Haifa pipeline.

Learned by me after 2004:

- Mike Makovsky had possible linkages to the Israeli Foreign Service.
- During the late 1990s, Dr. Ahmed Chalabi promised the Israel lobby that he would provide Iraqi oil to Israel over the Haifa pipeline if the lobby would support him as the first prime minister after Saddam.
- The Israeli infrastructure minister, Joseph Paritzky, publicly announced on March 31, 2003, his intention to reopen the Kirkuk-to-Haifa pipeline and receive Iraqi oil before the end of 2003.
- The Israeli finance minister, Benjamin Netanyahu, hyped investors in London in June 2003 so that they would provide capital for rebuilding the Haifa pipeline.
- Secure crypto-communications were installed between Chalabi's offices in Baghdad and the office of the vice president where Scooter Libby was chief of staff.
- Scooter Libby made a small fortune as Marc Rich's lawyer over a sixteen-year period, during the 1980s and 1990s. Marc Rich was

known as the "King of Oil" because of his lucrative Middle East oil trading business, specifically the billions of dollars he made from a secret pipeline through Israel.

- A former CIA official stated that it had long been a dream of a powerful section of the people driving the administration of President George W. Bush and the war in Iraq to safeguard Israel's access to oil.

Taken individually, each of these facts was circumstantial and did not support an agenda case. However, when the dots were connected, the case became more than probable. It took me more than twelve years to realize why the people in Doug Feith's office were cheering so loud when Secretary Powell so convincingly stated the casus belli before the United Nations.

A discussion of each of these bullet points is appropriate. Let's take them in the order listed.

Senior political appointees at the Pentagon pushed for Dr. Ahmed Chalabi to be prime minister. Secretary Rumsfeld may not have officially supported Chalabi for Iraq's first post-Saddam prime minister position, but there was no doubt that his subordinates were officially and unofficially supporting Chalabi. As I stated in an earlier chapter, my Kuwait roommate was a retired navy SEAL, Terry Sullivan. He and Dr. Harold Rhode were sent to Kuwait by Doug Feith to link up with Chalabi's people.

Mike Makovsky informed the EIPG team during prewar planning that Chalabi would be Iraq's first post-Saddam prime minister.

Retired Lieutenant General Jay Garner was the first civilian head of Iraq. Jay advised me in 2016 that Doug Feith met with him just before his departure for Kuwait. Feith suggested that Garner turn Iraq over to Chalabi shortly after Baghdad fell.

Feith became livid in a telephone conversation with Garner a few days later after Jay answered questions at a press conference before departing the Pentagon for Kuwait. His answer to a question that made Feith so angry: Jay said that he and the US government had not chosen Chalabi or any other specific Iraqi to lead Iraq after the fall of Saddam.[9]

It did not take long for Chalabi to walk away from his promise to the neocons of opening a pipeline to Haifa. Anyone with an understanding of Iraqi society would know that Chalabi could not deliver on this guarantee, but the neocons had a strong motivation to believe his prewar promises. His duping of the neocons became public information in 2004. An article in Salon.com from 2004, "How Ahmed Chalabi Conned the Neocons," identifies the details of how he deceived them. It includes a quotation from Doug Feith's former law partner at Feith and Zell, PC, who was trying to develop his Israeli law practice with Chalabi's promises. Marc Zell continued to be a partner in the firm that Feith left in 2001 to take the Pentagon job. "He said he would end Iraq's boycott of trade with Israel, and would allow Israeli companies to do business there. He said the new Iraqi government would agree to rebuild the pipeline from Mosul in the northern Iraqi oil fields to Haifa. But Chalabi has delivered on none of them." L. Marc Zell went on to say, "Ahmed Chalabi is a treacherous, spineless turncoat."[10]

The neocons were too hard on their Iraqi partner. He did the only thing he could do in 2003 to secretly provide oil to Israel. As I stated earlier in the book, Chalabi was successful at selling Iraqi oil to the oil trading company that had supplied most of Israel's oil since 1973, Glencore.[11] The SOMO director general in mid-2003, Mohammed Al Jibouri, only sold oil to end users, refineries. He refused to sell to oil traders during CPA. Chalabi ordered Al Jibouri to sell Iraqi oil to Glencore, and Al Jibouri refused. In October 2003, Chalabi instructed oil minister Ibrahim Bahr Al Ulloum to replace the SOMO director general with an Iraqi that would sell oil to Glencore.[12] Al Jibouri was removed.

The only deliberate oil infrastructure target during military operations was the K-3 pump station that pumped oil to Syria. The US military hit the K-3 pump station in Anbar Province during the invasion. That pump station was pumping crude oil to ports in Syria prior to the invasion. While K-3 was still burning from the attack, Secretary Rumsfeld stated to the press that the pump station was destroyed in order to penalize Syria for enabling Saddam to bypass sanctions prior to the invasion.[13] Such an explanation made no sense. It was not logical. Syria's future benefit under a US occupation or new Iraqi government would only be the tolls received from those exports. Those tolls represented less than 5 percent of the value of the crude exported. The other 95 percent of that value was for Iraq. The destruction of Iraq's K-3 pump station hurt Iraq much more than Syria. Was there a more credible explanation for destroying the pump station?

An action memo from Feith to Rumsfeld on January 16, 2003, specifically directed the military to "Keep the pipeline open only if Syria cooperates on issues of interest to US/Coalition."[14] The memo was declassified in 2015. Paul Wolfowitz verbally reversed those written orders a month after issuance and after Secretary Powell made his WMD presentation to the United Nations. Wolfowitz gave his direction to General Franks during a video teleconference that took place a couple of weeks before the Iraq invasion. I was present in the meeting.

The following quotation in the London press shortly after the K-3 destruction identified the motivation: "Its [the Haifa pipeline's] resurrection would transform economic power in the region, bringing revenue to the new

US-dominated Iraq, cutting out Syria and solving Israel's energy crisis at a stroke."[15]

Our mission was to restore the oil sector to prewar levels, but Phil Carroll was ordered by the Pentagon not to reconstruct the destroyed K-3 pump station. A credible and logical explanation for the destruction of K-3 was to remove any competition to future Haifa pipeline expansions.

Mike Makovsky contacted me from the Pentagon during the summer of 2003 to get a status on the Haifa pipeline. My first exposure to the existence of the old Haifa pipeline was when Makovsky phoned from the Pentagon to inquire about its status. He labeled his inquiry to me as a top-priority project from the civilian leadership at the Pentagon. He hung up the phone on me when I realized his inquiry was about a very sensitive pipeline to Israel and requested that he submit his request in writing. This incident was covered in detail in an earlier chapter. In Steve Coll's book *Private Empire*, Mike claimed that he was responding to an Israeli inquiry about the pipeline.[16]

Mike Makovsky had links to the Israeli government. It was revealed in 2005 that Makovsky had left the United States for Israel in 1989 to join the Israeli Foreign Service.[17] In December 2005, I asked Mike about the reports of his time in the Israel Defense Forces (IDF) and his response was that he would neither confirm nor deny these allegations. Phil Carroll advised me in 2005 that Makovsky received his top secret security clearance only after Doug Feith trumped the challenges of the Pentagon security group. Mike Mobbs said that Makovsky was hired at the Pentagon in 2002 because of his AIPAC contacts.[18] AIPAC was involved in an espionage scandal in Doug Feith's group in 2005.[19] This raised my suspicions that Makovsky was possibly some type of an agent for the Israeli government. It also raised my suspicions about why an Israeli agent with little to no oil and gas experience was assigned as the Pentagon's Iraq oil contact.

During the 1990s, Dr. Ahmed Chalabi promised the Israel lobby that he would provide Iraqi oil to Israel over the Haifa pipeline if the lobby would support him as the first prime minister after Saddam. Chalabi made a rousing speech to the Jewish Institute for National Security Affairs (JINSA) in June 1997. JINSA is a right-wing, pro-Israel advocacy group, part of the powerful Israel lobby. During that speech, Chalabi proposed expanding economic ties between Iraq and Israel by restoring the Kirkuk-to-Haifa pipeline.[20] Ten years after the Iraq invasion, Mike Makovsky was appointed CEO of JINSA.[21]

• • •

The Israeli infrastructure minister, Joseph Paritzky, publicly announced on March 31, 2003, his intention to reopen the Kirkuk-to-Haifa pipeline and receive Iraqi oil before the end of 2003. While the coalition forces were still fighting their way to Baghdad, Minister Paritzky was making public announcements about how Israel would economically benefit from the US war. The Israeli news service, Haaretz, quotes Paritzky on March 31, 2003, as saying that the reopened Kirkuk-to-Haifa pipeline would cut Israel's energy bill drastically, by more than 25 percent. In 2003, Israel was very dependent on imported oil to meet its energy needs. Israel was getting most of its oil from Russia at the time and Paritzky felt that Israel was paying a premium for that oil.[22]

Paritzky was quoted in *Haaretz* later in the summer of 2003 as working closely with sources inside the Pentagon. A telegram from the Pentagon in August to the Israeli government requested that Israel check the possibility of pumping oil from Iraq to Haifa.[23]

The Israeli finance minister hyped investors in London in June 2003 so that they would provide capital for rebuilding the Haifa pipeline. Paritzky was not the only member of Israel's cabinet in 2003 publicly making statements about getting Iraqi oil. The Israeli finance minister in 2003 was Benjamin Netanyahu. He was in London in June 2003 promoting the Haifa pipeline to potential investors. "It won't be long when you will see Iraqi oil flowing to Haifa. It is not a pipe dream," Netanyahu told the investor group.[24]

Secure crypto-communications were installed between Ahmed Chalabi's offices in Baghdad and the office of the vice president, where Scooter Libby was chief of staff. A STU-III looks like an old telephone from the 1980s. However, it is a highly technical phone that enables encrypted secure voice transmission. One was installed in Chalabi's Iraqi National Congress (INC) headquarters for communications to the vice president's offices in DC. An army colonel was sent to support Chalabi with this secure communication. Unfortunately, that colonel had to be evacuated from Iraq on emergency leave. Rather than sending a new officer to support this effort, CENTCOM sent a colonel to retrieve the STU-III. The Chalabi headquarters did not want to give up the instrument. It took several weeks to retrieve it. This communication link was a secure line back to the office of the vice president in Washington, DC. Scooter Libby was the vice president's chief of staff.

Scooter Libby made a small fortune as the lawyer of Marc Rich back in the 1980s and 1990s.[25] Marc Rich was well known within the oil trading business as one of the most successful Middle East oil traders. He made billions of dollars over a twenty-year period from a secret pipeline through Israel. That pipeline transported Iranian oil through Israel to the Mediterranean. Rich transported Iranian oil while Iran held American hostages and during the following years when Iran was under US sanctions. Such transactions after the Iranian revolution would have been illegal for a US company, but his company was registered in Switzerland.[26]

Rich was raised in New York City and joined the commodities trading group Philips Brothers. He eventually moved to Europe with Philips but parted ways over a disagreement about the size of his bonus. He started his own company, Marc Rich & Co, AG, in Switzerland. This company would become Glencore International, a well-known oil-trading group. His specialty was trading oil with countries under US sanctions. His biggest customers included Iran and South Africa during sanctions periods. Others included Cuba, Marxist Angola, Sandinistas, Romania, and others under US sanctions.[27]

The Internal Revenue Service investigated Rich's multibillion-dollar business during the 1980s for evading over $100 million in back taxes. More specifically, the US attorney for the Southern District of New York, Rudolph W. Giuliani, prosecuted the case against Rich. Marc Rich did not return to the United States for his court appearance and the FBI listed him as one of its top international fugitives. The title of the Interpol arrest warrant in 1985 was "Wanted International Criminal" and the charges included income tax evasion, racketeering, and trading with the enemy.[28]

Rich became well known in US political circles. President Clinton signed a pardon for Rich and his partner, Pincus Green, late on January 19, 2001, Clinton's last day in office, a controversial end-of-term pardon. William Safire, a Pulitzer Prize–winning columnist for the *New York Times* called the pardon "the most flagrant abuse of the presidential pardon in U.S. history."[29]

Scooter Libby testified before Congress on behalf of Marc Rich.[30]

To better understand why Marc Rich was a key figure with the Haifa pipeline agenda, it is important to understand not only his relationship to Libby, but also his oil-trading history with Israel. Rich had a close relationship with several Israeli prime ministers. His company, Glencore, was an important supplier of oil to Israel for decades. Marc Rich was close to the intelligence arm of Israel, the Mossad. The Mossad referred to people like Rich as "Sayan," the Hebrew word for helper.[31]

Libby was also close to the Israeli government. The former British foreign secretary from 2001 to 2006, Jack Straw, said of Libby: "It's a toss-up whether [he] is working for the Israelis or the Americans on any given day."[32]

There was a secret Israeli pipeline between the Red Sea and the Mediterranean. The pipeline starts at the Israeli port of Eilat, adjacent to the

Jordanian port of Aqaba. It is 160 miles long and forty-two inches in diameter and ends at the Israeli port of Ashkelon. This pipeline through Israel carried a sizeable amount of Iranian crude oil and accounted for most of the wealth that Marc Rich accumulated. It flowed before and after the United States imposed sanctions on Iran. Rich and his lawyers kept the transactions confidential so that no one knew Iranian crude oil was being delivered to Israel. It was a win-win deal for everyone involved. The Iranians had a customer for their oil. The Israelis received access to inexpensive Iranian crude oil for their refineries. And Marc Rich made billions of dollars. Israel received 60 to 90 percent of its oil needs from Iran between 1973 and 1994.[33]

An article in the *Guardian* in April 2003 spoke of the history of key people in the Bush Administration and their dream of getting Iraqi oil for Israel. It quoted a former senior CIA official: "It has long been a dream of a powerful section of the people now driving this administration [of President George W. Bush] and the war in Iraq to safeguard Israel's energy supply as well as that of the United States."[34]

An effort was made by the US government in 1983 to move Iraqi crude oil to the port of Aqaba, adjacent to the southern end of the Israeli pipeline. Such volumes could have replaced Iranian crude oil on the secret Israeli pipeline. Donald Rumsfeld was CEO of G. D. Searle & Co. in the early 1980s. Rumsfeld was appointed as a presidential envoy by President Reagan to visit many Middle East countries. He visited Saddam in Baghdad in December 1983 and in March 1984. The Iran-Iraq War had taken a toll on Saddam's ability to export his oil. Iran had shut down Iraq's oil exports through the Basra ports. Syria was Iran's ally at the time. They shut down Iraq's export pipeline through Syria. Saddam was totally dependent on exports through Turkey to the Mediterranean. One of Rumsfeld's agenda items was to help Saddam transport Iraq's crude through a different route. Rumsfeld proposed an export pipeline through Jordan to Aqaba. The ports of Aqaba and Eilat are adjacent to each other at the northeast end of the Gulf of Aqaba. Such a pipeline would have enabled Israel to access Iraqi crude oil for its secret pipeline rather than Iranian crude oil, which was under US sanctions.[35]

An initiative to send a private citizen as a presidential envoy to the Middle East would have required significant coordination through the State Department. Paul Wolfowitz was the director of policy planning at DOS in 1981 and 1982. He was moved to assistant secretary of state for East Asian and Pacific affairs in 1983. Wolfowitz brought Scooter Libby into DOS to work for him during this period. Paul Wolfowitz had been Scooter Libby's mentor since teaching Libby at Yale in the 1970s.[36]

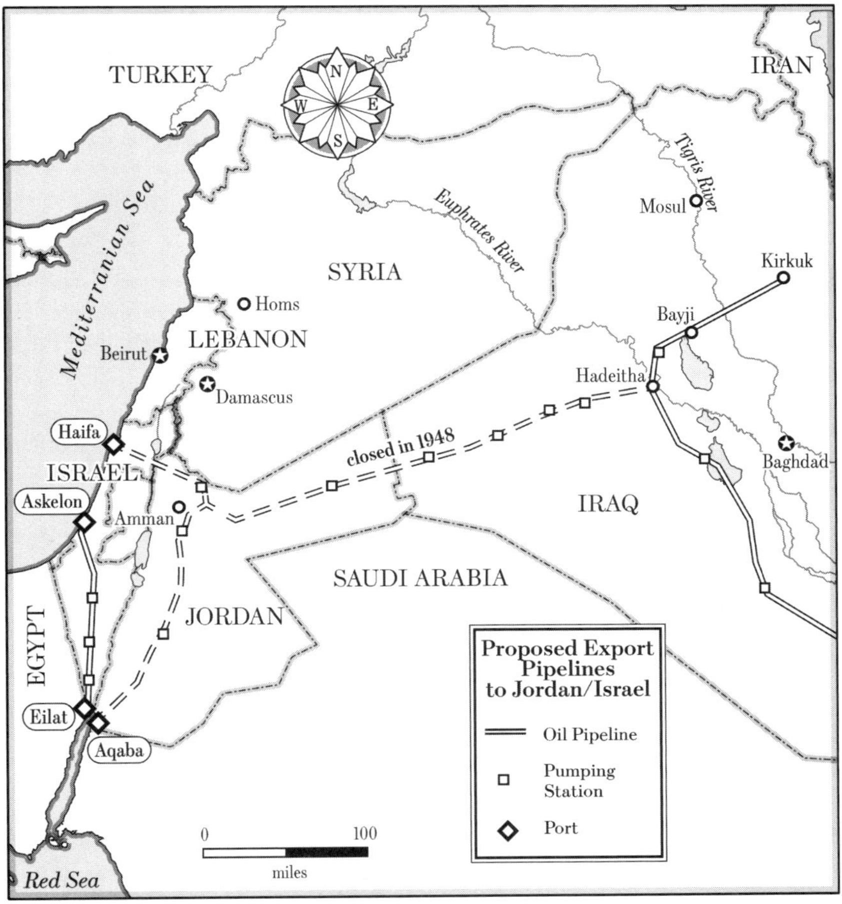

Proposed Export Pipelines to Jordan/Israel as of 2003.

• • •

Billions of US dollars were at stake for a Haifa pipeline. The suggested pipeline size would have transported $30 to $50 billion per year of Iraqi oil through Israel. Was it possible that the neocons were motivated by the opportunity for financial gain with an Iraqi pipeline to Haifa? Would a pipeline to Aqaba have been a fallback position if rebuilding the Haifa pipeline in 2003 were not acceptable to Iraq? Would many of Friedman's twenty-five neoconservatives get access to the economic gains from a pipeline to Haifa or Aqaba? I think it is naïve to believe that greed did not play a role in their motivation to push the Iraq War. Dr. Ahmed Chalabi made prewar promises about opening trade between Iraq and Israel. He promised to open a pipeline to Haifa for exporting millions of barrels per day of crude oil. How could the highly educated and brilliant neocons have fallen so easily for the lies of a known and unscrupulous opportunist like Dr. Ahmed Chalabi? How were they so easily blinded to what was obvious to the CIA and State? Does greed explain this blindness?

The costs to the United States for invading Iraq were huge. There were 4,489 sons and daughters killed, 32,223 troops injured (not including PTSD), at least 134,000 Iraqi civilians killed, 150 reporters killed, 2.8 million people who either remain internally displaced or have fled Iraq, and a financial cost of more than $2 trillion.[37]

Did my country incur all of these costs because of a couple dozen neocons? Did a network of highly educated, right-wing, pro-Israel US government senior political appointees develop a plan to enrich themselves and their colleagues? Did they bring in staffers with a similar ideology to help execute such a plan? Some staffers, like Mike Makovsky, arrived in 2002 with direct ties to the Israeli government.

Furthermore, this group of neocons is still giving speeches, promoting the next Middle East war, lecturing at universities, appearing as guests on television political shows, and receiving honorariums. They feel no shame. No remorse. They never attended a Saturday BUA or listened to the chaplain attempt to explain the loss of a young soldier. They missed that completely. What haunts me is that I may have been misled into believing that WMD were the noble reason for war when it was just an excuse for their war. Was I an unknowing participant in an oil scheme?

What troubles me are the grieving parents, or spouse, or children of that soldier who made the ultimate sacrifice. Or the caregiver of that disabled soldier who can no longer care for himself or herself. It is essential that the survivors understand the noble cause of the military profession. It is a principled cause of duty to country and a love of country. There is something about raising your right hand and swearing to defend the Constitution of our country against our enemies. This is a noble cause. It is a cause that brings patriotic Americans together on a military team. Being part of a military team engaged in battle is an event that few experience. The brotherhood of those team members is extremely close. Studies have shown that soldiers are not motivated during battle because of their president, or congressmen, or other political leaders. No, soldiers fight for one another. It is the honor of their profession to die for their fellow soldiers. It is a creed among brothers and sisters. It is their creed of duty, honor, and country.

America must be vigilant in the future. We cannot allow another group, such as the neocons, to push our country into a costly and unnecessary Middle East war as they did in Iraq.

Epilogue

We don't blame President Bush or Obama. We blame Maliki.
—Several Iraqi citizens in June 2015

Iraq was producing and exporting oil at record levels in late 2015. In spite of the poor security situation created by ISIS in 2014 and 2015, Iraq increased oil production. Can Iraq maintain these record levels? Will Iraq further increase production? Why did the Kurds have to ship most of their oil to Israel? What was happening in Iraq's refining sector after ISIS took control of the Bayji Refinery? How were the depressed global oil prices impacting Iraq's oil sector? What was the level of investment by the US oil companies in the southern oil fields at the end of 2015? Let's take a look at each of these questions.

It was May 2015 and I found myself back in Baghdad working on a consulting engagement for the US government. It was my first trip back since departing in September 2011.

Security inside Baghdad seemed worse than it was in 2011 but not as bad as in 2006. The Iraqi army had moved several thousand more soldiers into Baghdad and set up checkpoints throughout. Suicide attacks in Baghdad would not start in high numbers until shortly after I left in June.

Basra security was fine. The hotels were all booked. There were traffic jams throughout the city. The airport was busy with foreign oil workers entering and departing. The only security issue with Basra was that many of the soldiers sent to Baghdad came from Basra. So, criminal activity was increasing.

It was good to visit with many of my old Iraqi friends and I made a few new ones. Although not part of my engagement, I asked many, "Who do you blame more, President Bush or Obama, for your current poor state of security?" That question was being debated on all the political programs on television in the United States. So, I was curious about what Iraqis thought. The answers somewhat surprised me.

During my previous time in Iraq, the Iraqis did not hesitate to blame their problems on the Americans. We were blamed for a lot. Some blame was warranted, most not. This time, the answer was consistent. They all blamed former Prime Minister Maliki and his sectarian policies. No one blamed Bush or Obama. That was encouraging. The Iraqis owned their problem. Could they fix it?

The oil sector needs to operate at a high level to keep the economy moving and give the government a chance to fix things. Any discussion about the oil sector should first consider security and its negative impacts since 2003. Every time the oil sector experienced positive momentum, the progress was overcome by events due to poor security.

Iraq developed an excellent Integrated National Energy Strategy (INES) that was approved by the government in 2012. It outlined a way forward that included target objectives for the energy sector. The ISIS attacks of 2014 put many of those objectives at risk. For example, the target production of nine million barrels per day by 2020 requires another 4.5 million barrels per day above the 2015 levels. No one currently believes that nine million barrels per day is achievable by 2020. However, a production number of about six million barrels per day might be achievable if prices increase from their depressed 2015 levels and security improves.

Upstream

Production and exports have made significant progress since 2011. The December 2015 production number of 4.4 million barrels per day included both the Baghdad-controlled oil fields in the south and the mix of Baghdad- and KRG-controlled fields in the north. The exports in the north could not use the old pipeline from Kirkuk to Bayji and then north to Turkey. ISIS controlled much of the route of that pipeline. Fortunately, KRG had installed a large pipeline before ISIS took control in 2014. The KRG pipeline was located in KRG-controlled territory and connected to the Iraq-Turkey export pipeline close to the Turkish border at the Feshkhabur pump station. KRG exported both KRG-controlled production and production from Baghdad's federal North Oil Company (NOC) oil fields through its pipeline.

Domestic crude oil consumption for refineries and power plants was about 0.6 million barrels per day in 2015. KRG accounted for 0.2 million barrels per day of that estimate.

Exports increased more than one million barrels per day over the twelve months of 2015. Such a rate of increase will slow significantly over the next few years. Besides security, the most significant obstacle will be the lack of financial resources to fund further development. Falling oil prices have decreased the cash available for all government expenditures, including oil field development.

On-time contract payments to the international oil companies (IOCs) will be important for further production increases. The IOCs required immediate cost recovery to achieve the economic guarantees of their fixed contracts. Delinquent payments would discourage oil field development by the IOCs. Low remuneration fees were negotiated with the IOCs in return for immediate cost recovery from Iraq. Shahristani's bid round contracts worked well during periods of high oil prices and relatively low demands on Iraq's cash. The contracts were negotiated when prices were around $75 per barrel. Those contracts were terrific when the price went above $100 per barrel. As prices fell to under $40 per barrel, his contracts became a financial burden to Iraq and payments to oil companies were delayed.

The annual development costs for the IOC contracts had been running at close to $20 billion per year before 2015. The government easily handled this cost when crude oil prices were $100 per barrel and the government budget was $100 billion per year. However, the government could not meet the burden of IOC payments with the lower oil prices. There were too many higher-priority demands on the government budget. Funding the war against ISIS was the biggest.

The contracts with the IOCs led to a significant increase in southern production after 2010. The IOCs introduced new technologies, improved management skills, developed human resources, much improved the focus on health and safety, improved environmental practices, and did other positive things for Iraq. The Iraqi oil professionals recognized the benefits that the IOCs brought to Iraq, but they also recognized that Iraq could not live with the current contract structure and the current low oil prices.[1]

Contracts were renegotiated in 2013. Production plateaus were reduced, the contract life was extended by five years, and the economics of the IOCs' projects were kept constant during the first renegotiation. Were additional renegotiations needed to recognize the realities of the fixed-fee service contracts in a low-price environment? Yes. The challenge was how to accomplish this within the parameters of fixed-fee service contracts for Iraq and still guarantee economics for the private sector oil companies. The first renegotiation accomplished this, but future amendments became very difficult. The only two American oil companies with contracts in southern Iraq, ExxonMobil and Occidental, have not been enthusiastic with their investments. Occidental withdrew from its Zubair contract at the end of 2015.[2] As previously mentioned, ExxonMobil substantially decreased their investment in the West Qurna 1 contract, from 80 percent to 25 percent, in 2013.[3]

There were also physical bottlenecks limiting production in the south. Southern exports averaged close to 3.4 million barrels per day at the end of 2015 (see Table E.1). These exports were twice the levels of 2011. The old export infrastructure system of ABOT/KAAOT offshore platforms histori-

Table E.1 South and Central Oil Production by Oil Field

Field	*Consortium Operator*	*Original Production Target (PPT) (barrels/day)*	*Production Avg. 2015 (barrels/day)*	*Production Base 2000 (barrels/day)*
Ahdab	CNPC	140,000	135,000	—
1st Bid Round				
Rumaila	BP	2,850,000	1,300,000	1,066,000
West Qurna (Ph 1)	ExxonMobil	2,325,000	380,000	244,000
Zubair	ENI	1,200,000	350,000	183,000
Maysan oil fields	CNOOC	450,000	130,000	88,000
2nd Bid Round				
Majnoon	Shell	1,800,000	200,000	45,000
West Qurna (Ph 2)	Lukoil	1,800,000	450,000	—
Halfaya	CNPC	535,000	200,000	3,000
Garraf	Petronas	230,000	100,000	—
Badra	Gazprom	170,000	45,000	—
Other Fields	Oil Ministry Companies	—	240,000	—
Total Production		11,500,000	3,530,000	1,629,000

Data Source: EIA and MoO presentation to Parliament, December 13, 2015

cally peaked at a ceiling of about 1.7 million barrels per day. The additional 1.7 million barrels per day were achieved after the commissioning of the single-point mooring (SPM) export system.

There were three SPMs operating in December 2015 at a rate of about 550,000 barrels per day for each SPM. They were operating at about 60 percent of their capacity. Additional volumes through the SPMs required higher onshore pumping capacity, and that was scheduled for 2018. A fourth SPM was awaiting the removal of a shipwreck before it could be commissioned. This fourth SPM should immediately increase the physical export capacity above 3.7 million barrels per day out of the south. A risk to this export volume was the continued use of the old subsea pipelines to ABOT. No one knew why those old decrepit pipelines had not yet failed.

The other significant physical bottleneck was the water-injection project discussed in chapter 18. Without a water project, several of the southern oil fields would soon approach their production ceilings. The oil ministry contacted a couple of the IOCs for help with this project in 2015, but the project was longer-term and would require extensive financial resources. It was a project that would probably be delayed until oil prices recovered.

Could Iraq achieve the production of six million barrels per day by 2020? This would require additional exports of about two million barrels per day. The southern export infrastructure could handle about 5.5 million barrels a day by 2020 if two projects are completed. Completion of the Al Fao terminal project in 2018 should add an additional 1.5 to 1.8 million barrels per day. The higher pumping capacity from Al Fao should allow optimization of the SPMs. The completion of the Japanese Sealine project should provide more reliable piping to ABOT/KAAOT and an additional SPM.

Production volumes from the IOCs should increase within the current contractual targets. The ten operational contracts at the end of 2015 still had more than 4.5 million barrels per day of incremental production under their amended contract plateaus. It is in the interest of the IOCs to ramp up capacity. It is also their contractual obligation. Continued low oil prices that place cash flow constraints on Iraq provide the biggest risk to the IOCs meeting their contractual plateaus over the near term.[4]

Iraq paid out almost $40 billion to the IOCs for the years 2011, 2012, and 2013. This included $3 billion of remuneration fees (profit). There was another $10 billion for 2014 that did not get paid until 2015. As stated earlier, the government had higher priorities for its limited cash revenues—the war against ISIS.[5]

Kurdistan Regional Government (KRG)

KRG's production averaged about 0.8 million barrels a day at the end of 2015 (see Table E.2). About 0.6 million barrels per day were exported through the Iraq-Turkey pipeline to Ceyhan, Turkey. When these exports were combined with Basra exports, Iraq's total exports reached close to four million barrels per day, the second highest in OPEC behind Saudi Arabia.

These KRG exports had a much higher risk attached to them than the Basra exports. The fight against ISIS was much closer to the KRG oil fields, and the export pipeline was even closer to the fight. But that was not KRG's biggest threat. KRG's exports were treated as smuggled oil by the Baghdad government. It could not sell its oil in the global markets without the threat of seizure by Baghdad. This was because of the ongoing constitutional disagreement between Baghdad and KRG. Baghdad's position was that only the central government could legally sell Iraqi oil, while KRG's position was that it had the authority to sell its oil produced in KRG.

The oil tanker *United Kalavryta*, carrying one million barrels of KRG crude oil, tried to unload its cargo at a Texas port in July 2014. The Baghdad oil ministry sued in US court for seizure of the cargo. The ship remained

Table E.2 KRG-Controlled Oil Field Production

KRG-Controlled Oil Fields	Operator	Production (barrels/day)
Eastern Kirkuk Fields	North Oil Company	220,000
Bai Hassan/Avana	KAR Group	220,000
Kurmala Dome	KAR Group	103,000
Taq Taq	TTOPCO	85,000
Tawke	DNO	128,000
Ain Zalah	NOC – KRG Controlled	8,000
Demir Dagh	Oryx Petroleum	0
Shalkan	Gulf Keystone	35,000
Sarsang	HCN Energy	2,000
Sarqala	Western Zagros	5,000
Khor Mor	Pearl Petroleum	14,100
(Source: Iraq Oil Report)		820,100

outside the jurisdiction of the court until appeals could be heard so that physical seizure could not take place. *United Kalavryta* sailed away from Texas in January 2015 when it was obvious that the Baghdad government would prevail in the seizure action. The ship sailed to the Israeli port of Ashkelon on the Mediterranean, where it was unloaded.[6]

More than half of the KRG exports were offloaded in Israel in 2015. Vessel transponders were shut off while the ship moved into port to offload its cargo. Then the transponder was turned back on after the ship returned to sea, and tanker-monitoring systems picked up their movement. Baghdad still refused to recognize the government of Israel, so it did not attempt a seizure action on KRG cargoes offloading Iraqi crude oil in Israeli ports. This was a nice windfall for the energy industry in Israel. The Israeli refineries were receiving more oil than they could ever process and the port loading facilities were reexporting the excess oil.[7]

The irony of this situation should not go unnoticed. Ahmed Chalabi failed to deliver oil to Israel through the Haifa pipeline, but the Kurds are now delivering Iraqi oil to Israel.

The remaining KRG cargoes not destined for Israeli ports were secretly trans-shipped at sea with transponders turned off to hide the movement of the oil. The smuggling effort of KRG was the largest oil-smuggling operation in the world at the time and may well be the largest in history.

All cargoes were sold to international oil traders at the port of Ceyhan. Transferring custody at Ceyhan enabled KRG to deny any direct involvement in the actions being taken and thereby to avoid seizure by the Baghdad government. The primary traders were Vitol and Trafigura, both headquartered in the Netherlands. The financial payment terms of this arrangement were prepayment to KRG's accounts. The traders committed about $850 million per month in 2015 in exchange for the applicable volumes to cover their cash. Such smuggling activity only drove up the costs of handling the KRG oil and the corresponding discounts to the trading companies. The resulting netback price to KRG was significantly discounted below global market prices.[8]

The eastern Kirkuk fields were the only fields operated by the North Oil Company (NOC) in 2015. Exports from all of these fields were for the accounts of KRG at the end of 2015.

Refining Sector

Iraq remained significantly short of refining capacity for providing petroleum products to the country. Expensive imports of gasoline and diesel made up for the shortages in 2015. These imports were another significant drain on the country's limited cash flow, especially because of the subsidized domestic fuels market.

Since 2004, the Ministry of Oil has tried to find an international partner for the refining sector. The larger international oil companies were not interested in investing in Iraq's refining sector unless the investment was linked to an upstream deal. The ministry attempted to implement such a linked deal with the Nasiriya Integrated Project (NIP), close to the city of Nasiriya. Two significant factors hurt the Nasiriya project. First, the proposed site was about 250 kilometers from the seaports, making it totally dependent on the domestic market. Refining sector investors like to be located close to port facilities to enable sales to the global energy markets. Second, Iraq had a program for subsidizing the sale of oil products. It needed to reform this distribution system and gradually remove subsidies.

The loss of the Bayji Refinery in 2014 to ISIS was a significant loss of refining capacity for Iraq. This loss removed about half of the gasoline and diesel production in Iraq. The oil ministry sent engineers to inspect Bayji in October 2015, just after its liberation. Their assessment indicated that damage was largely confined to storage tanks and one of the three process trains. Limited operations could return to the refinery in less than a year. This was a pleasant surprise to the oil ministry.

The good news did not last long. Reports surfaced in the press at the end of 2015 indicating that severe looting had occurred at Bayji late in the year. This looting did not take place during the occupation by ISIS but during the occupation by the Shia militia inside the refinery in November and December 2015. The Shia militia refused to allow engineers to enter the refinery while the looting operation was being executed. The return of Bayji to partial operations may now take much longer, if it ever happens.

The Doura and Basra refineries continued operating at their historical levels. There were ongoing upgrade projects, but these projects were moving at a very slow pace. The shortage of cash precluded any significant capital investment in these two refineries.

The Karbala Refinery project was the only solid project in sight over the next few years. This was a new refinery of 140,000 barrels per day located southwest of Baghdad near Karbala along the Strategic pipeline. Contracts were awarded to two large international companies. The estimated cost of this refinery was $6 billion and it was part of SCOP's portfolio of megaprojects. Funding for this project was coming from the oil ministry's capital budget. It was another demand on the country's limited cash flow.[9]

Final Comments

The oil sector has been the one successful constant in Iraq since 2003. Production stopped during hostilities in March 2003. The oil flowed again on April 23, 2003, and the oil has not stopped flowing in Iraq since then. It has fluctuated, but it has never stopped. The next ten years should see more success as long as the international oil companies remain engaged.

The oil fields of southern Iraq are some of the largest in the world. The cost to produce a barrel of oil out of these southern fields is among the lowest in the world. Economics favor Iraq. Current exports of 3.4 million barrels per day from southern production should be twice as high in 2025.

Except for continued low oil prices or the departure of the international oil companies from the south, it is difficult to forecast a scenario where Iraq will not be exporting more than six million barrels per day from the southern oil fields by 2025. The security issues were primarily localized to areas north of the Shia city of Najef. The southern oil fields were excluded from ISIS violence. Southern Iraq should grow its oil sector over the next decade.

The future of the northern part of the country is not clear. What happens to KRG and the Sunni provinces of the north after ISIS is defeated? Will KRG remain part of Iraq? Does the Sunni north form its own regional government? These are all political issues that will impact the oil sector in the north.

Notes

Chapter 1. American Oil Minister

1. Michael Kelly and Jeffrey Ingersoll, "The Staggering Cost of the Last Decades US War in Iraq—in Numbers," *Business Insider*, June 20, 2014.

Chapter 2. Contingency Planning for the Oil Sector

1. Interview with Michael Mobbs on March 13, 2015, by Gary Vogler.
2. Ibid.
3. Ibid.
4. Ibid.
5. Follow-up interview with Michael Mobbs on December 10, 2015, by Gary Vogler.
6. Aram Roston, *The Man Who Pushed America to War* (New York: Nation, 2008), 193–199.
7. Donald Rumsfeld, *Known and Unknown: A Memoir* (New York: Penguin, 2011), 488.
8. Jack Herman, "A Whole New Ballgame Overseas," *St. Louis Post*, February 20, 1989.
9. David Johnston, "Pentagon Analyst Gets 12 Years for Disclosing Data," *New York Times*, January 20, 2006.
10. Steve Coll, *Private Empire: ExxonMobil and American Power* (New York: Penguin, 2012), 221.
11. Interview with Clarke Turner on March 6, 2015, by Gary Vogler.
12. Interview with B. J. Ramos on November 10, 2014, by Gary Vogler.
13. Gary Vogler, *Lessons Learned—The Iraq Energy Sector* (Fairfax: Howitzer, 2016), appendix.
14. Ibid.
15. Dan Murphy, "Iraq War: Predictions Made, and Results," *Christian Science Monitor*, December 22, 2011.
16. Vogler, *Lessons Learned*, appendix.
17. Ibid.
18. Ibid.

Chapter 3. ORHA Start-Up

1. Gordon Rudd, *Reconstructing Iraq: Regime Change, Jay Garner, and the ORHA Story* (Lawrence: University Press of Kansas, 2011), 95.

2. Ibid., 145.

3. Ibid., 148.

4. Interview with Jay Garner on March 14, 2016, by Gary Vogler.

5. Rudd, *Reconstructing Iraq*, 299.

6. Interview with Jay Garner on March 14, 2016, by Gary Vogler.

7. Ibid.

8. Rudd, *Reconstructing Iraq*, 142.

Chapter 4. Iraqi Oil Sector

1. Dr. Ferruh Demirmen, "Oil in Iraq: The Byzantine Beginnings," *Global Policy Forum*, April 25, 2003.

2. Ibid.

3. Phebe Marr, *The Modern History of Iraq* (Boulder: Westview, 1985), 202.

4. "Brief History," Organization of the Petroleum Exporting Countries, website, May 2016, www.opec.org/opec_web/en/about_us/24.htm.

5. Brian Levy, "World Oil Marketing in Transition," *International Organization* 36, no. 1 (Winter 1982): 113–133.

6. Luis de Sousa, "OPEC Quotas and Crude Oil Production," *The Oil Drum: Discussions about Energy and Our Future*, blog, January 31, 2011, www.theoildrum.com/node/7363.

7. Stanley Ramsey, "Crude Oil and Liquid Products Infrastructure Chart," unpublished mapping project from the author's private library, provided by Ramsey, October 2009.

8. Iraq Petroleum Company, *An Account of the Construction in the Years 1932 to 1934 of the Pipe-Line of the Iraq Petroleum Company, Limited, from Its Oilfield in the Vicinity of Kirkuk, Iraq, to the Mediterranean Ports of Haifa (Palestine) and Tripoli (Lebanon)* (Oakland, 1934). Summarized in an Internet article, accessed May 2016, http://en.wikipedia.org/wiki/Mosul–Haifa_oil_pipeline.

9. David L. Phillips, *Losing Iraq: Inside the Postwar Reconstruction Fiasco* (New York: Basic, 2005), 71.

10. Akiva Eldar, "Infrastructure Minister Paritzky Dreams of Iraqi Oil Flowing to Haifa," *Haaretz*, March 31, 2003.

11. Ed Vuillamy, "Israel Seeks Pipeline for Iraqi Oil," *Guardian*, April 19, 2003.

12. Stephen Scheer, "Netanyahu Says Iraq-Israel Oil Line Is No Pipe-Dream," *Reuters*, June 20, 2003.

13. Dr. Jeremy Goff and Saad Jassim, *The Kirkuk-Banias Pipeline*, Pipelines International, March 2011.

14. Stanley Ramsey, Crude Oil and Liquid Products Infrastructure Chart, unpublished mapping project from the author's private library, provided by Ramsey, October 2009.

15. M. S. Vassiliou, *The A to Z of the Petroleum Industry* (Plymouth, UK: Scarecrow, 2009), 278. See also "Iraq Pipeline through Saudi Arabia (IPSA)," *OpenOil*, modified January 30, 2013, http://wiki.openoil.net/Iraq_Pipeline_through_Saudi_Arabia_(IPSA).

Chapter 5. Adjusting the Oil Plan

1. Gordon Rudd, *Reconstructing Iraq: Regime Change, Jay Garner, and the ORHA Story* (Lawrence: University Press of Kansas, 2011), 161.

2. Ibid., 233.

3. Ibid., 161.

4. Interview with Doug Feith on January 27, 2005, by Gordon Rudd.

5. Interview with Ambassador Jerry Bremer on December 2, 2015, by Gary Vogler.

6. Email from Asri Moussa to Gary Vogler, July 10, 2016.

Chapter 6. Penalizing Syria

1. Gary Vogler, *Lessons Learned—The Iraq Energy Sector* (Fairfax: Howitzer, 2016), appendix.

2. Interview with Wayne Dudding on November 21, 2014, by Gary Vogler.

3. Ibid.

4. Ibid.

5. David Sanger and Thom Shanker, "A Nation at War: White House; Bush Says Regime in Iraq Is No More; Syria Is Penalized," *New York Times*, April 16, 2003.

6. Edmund Ghareeb with Beth Dougherty, *Historical Dictionary of Iraq* (Plymouth, UK: Scarecrow, 2013), 475. Also "Kirkuk-Banias Pipeline (IPC)," *OpenOil*, accessed April 2016, http://wiki.openoil.net/Kirkuk-Banias_Pipeline (IPC).

7. Ed Vuillamy, "Israel Seeks Pipeline for Iraqi Oil," *Guardian*, April 19, 2003.

Chapter 7. Finding Gold in Baghdad

1. Gordon Rudd, *Reconstructing Iraq: Regime Change, Jay Garner, and the ORHA Story* (Lawrence: University Press of Kansas, 2011), 299.

Chapter 8. The Calm before the Storm

1. Gary Vogler, *Lessons Learned—The Iraq Energy Sector* (Fairfax: Howitzer, 2016), appendix.

2. Ibid.

Chapter 9. The Long Hot Summer of 2003

1. Interview with Clarke Turner on March 6, 2015, by Gary Vogler.

2. Ibid.

Chapter 10. The Bizarre Top Priority

1. Jack Herman, "A Whole New Ballgame Overseas," *St. Louis Post*, February 20, 1989.

2. Steve Coll, *Private Empire: ExxonMobil and American Power* (New York: Penguin, 2012), 234.

3. Anton La Guardia, "Iraq-Israel Oil Pipeline 'To Reopen,'" *Telegraph*, June 21, 2003.

4. David L. Phillips, *Losing Iraq: Inside the Postwar Reconstruction Fiasco* (New York: Basic, 2005), 71.

5. Charles Grant, "Blair's Five Wars," *Prospect*, October 30, 2003.

6. Stephen Scheer, "Netanyahu Says Iraq-Israel Oil Line Is No Pipe-Dream," *Reuters*, June 20, 2003.

7. Ed Vuillamy, "Israel Seeks Pipeline for Iraqi Oil," *Guardian*, April 19, 2003.

8. Daniel Ammann, *The King of Oil: The Secret Lives of Marc Rich* (New York: St. Martin's, 2009), 65.

9. W. Patrick Lang, "Drinking the Kool-Aid," *Middle East Policy Council* 11, no. 2 (Summer 2004): 8.

10. Michael Gordon, "The Struggle for Iraq: Reconstruction; 101st Airborne Scores Success in Northern Iraq," *New York Times*, September 4, 2003.

11. Hugh Pope, "Iraq Adds Complexity to U.S., Syria Relations," *Wall Street Journal*, October 20, 2003.

12. Gordon Rudd, *Reconstructing Iraq: Regime Change, Jay Garner, and the ORHA Story* (Lawrence: University Press of Kansas, 2011), 148.

13. "Iraq Coalition Casualties: Fatalities by Year and Month," Iraq Coalition Casualty Count, March 2016, http://icasualties.org/Iraq/ByMonth.aspx.

Chapter 11. Task Force Shield

1. Interview with Tom O'Donnell on March 22, 2015, by Gary Vogler.

2. Rory McCarthey, "Suicide Raids on Oil Platforms Mark New Threat," *Guardian*, April 25, 2004.

3. Interview with Tom O'Donnell on March 22, 2015, by Gary Vogler.

4. Ibid.

5. Ibid.

6. Ibid.

7. Ibid.

8. Ibid.

Chapter 12. Changing of the Guard

1. Interview with Mohammed Al Jibouri on June 7, 2016, by Gary Vogler.

2. Daniel Ammann, *The King of Oil: The Secret Lives of Marc Rich* (New York: St. Martin's, 2009).

3. Email from DIA to Gary Vogler, August 27, 2003, Subj: SWD RIO Sitrep 27 Aug.

4. Interview with Rob McKee on March 23, 2015, by Gary Vogler.

5. Ibid.

6. Internet article, "Rebuilding Iraq: US Achievements through the Iraq Relief and Reconstruction Fund," US Department of State, February 2006, https://2001-2009.state.gov/p/nea/rls/rpt/60857.htm.

Chapter 13. CPA Closes Out

1. Interview with Rob McKee on March 23, 2015, by Gary Vogler.

2. Interview with Mike Stinson on March 23, 2015, by Gary Vogler.

3. Interview with Robert Maguire on August 18, 2015, by Gary Vogler.

4. Ibid.

5. Ibid.

6. Ibid.

7. Ibid.

8. "Iraq Coalition Causalities: Fatalities by Year and Month," Iraq Coalition Casualty Count, accessed March 2016, http://icasualties.org/Iraq/ByMonth.aspx.

9. *Reuters*, "DNO: Iraq Spat with Ex-US Diplomat Costs up to $92 million," *Ekurd Daily*, May 1, 2011.

10. "Iraq Coalition Causalities: Fatalities by Year and Month," Iraq Coalition Casualty Count, accessed March 2016, http://icasualties.org/Iraq/ByMonth.aspx.

11. Martin Cherrington, Kenneth Week, and Harold Anderson, "Directional Drilling For Gas Line Sets Two Records," *Oil & Gas Journal*, September 6, 1993. Also, "Directional Drilling," *PetroWiki*, modified June 26, 2015, http://petrowiki.org/Directional_drilling.

12. Interview with Mike Stinson on March 23, 2015, by Gary Vogler.

13. Gary Vogler and Kevin Ross, "Iraqis Mending Own Pipelines," *Oil & Gas Journal*, February 16, 2009.

Chapter 14. A Break between Tours

1. Michael Kelly and Jeffrey Ingersoll, "The Staggering Cost of the Last Decades US War in Iraq—in Numbers," *Business Insider*, June 20, 2014.

Chapter 15. Energy Fusion Cell

1. Interview with Major General Richard Cripwell on October 29, 2015, by Gary Vogler.

2. Ibid.

3. Ibid.

4. Ibid.

5. Ibid.

6. Ibid.
7. Wolf Blitzer, "CNN Wolf Blitzer Reports," CNN.com, December 8, 2004.
8. Interview with Pat MacArevey on November 12, 2015, by Gary Vogler.
9. Sean D. Naylor, "Close Call," *Military Times*, November 24, 2008.
10. Interview with Hamish McNinch on November 19, 2015, by Gary Vogler.

Chapter 16. The Bayji Refinery and the Surge

1. Interview with Major Christian Cook on October 23, 2015, by Gary Vogler.
2. Interview with Dr. Ali Obaidi on December 17, 2014, by Gary Vogler.
3. Ibid.
4. Email from General David Petraeus to Gary Vogler, November 11, 2015.
5. Interview with Dr. Ali Obaidi on December 17, 2014, by Gary Vogler.
6. Ibid.
7. Interview with Robert Carruthers on December 15, 2015, by Gary Vogler.
8. Interview with Dr. Ali Obaidi on December 17, 2014, by Gary Vogler.
9. Interview with Major Joe DaSilva on October 13, 2015, by Gary Vogler.
10. Ibid.
11. Ibid.
12. Interview with Major Christian Cook on October 23, 2015, by Gary Vogler.
13. Interview with Dr. Ali Obaidi on December 17, 2014, by Gary Vogler.
14. Interview with Major Joe DaSilva on October 13, 2015, by Gary Vogler.

Chapter 17. The Southern Export Project

1. Gary Vogler, "Iraq Crude Oil Export Expansion Heightens Country's Security," *Oil & Gas Journal*, May 2012.
2. Ibid.

Chapter 18. The Bid Rounds

1. Ruba Husari, "First Oil Bid Round Sputters," *Iraq Oil Forum*, June 30, 2009.
2. Ben Lando, "The Numbers: Gas Field Bidding Round Results," *Iraq Oil Report*, October 20, 2010.
3. These Iraqis should remain anonymous.

Chapter 20. What Haunts Me

1. Michael Kelly and Jeffrey Ingersoll, "The Staggering Cost of the Last Decades US War in Iraq—In Numbers," *Business Insider*, June 20, 2014.
2. Daniel Ammann, *The King of Oil: The Secret Lives of Marc Rich* (New York: St Martin's, 2009), 232.
3. "Quotes," Dwight D. Eisenhower Presidential Library, Museum, and Boyhood Home, October 2015, www.eisenhower.archives.gov/all_about_ike/quotes.html.

4. Ari Shavit, "White Man's Burden," *Haaretz*, April 3, 2003.

5. Kathleen Christison and Bill Christison, "The Bush Neocons and Israel," *Counterpunch*, September 6, 2004, www.counterpunch.org/2004/09/06/the-bush-neocons-and-israel/.

6. Muhammad Idrees Ahmad, *The Road to Iraq: The Making of a Neoconservative War* (Edinburgh: Edinburgh University Press, 2014), 12.

7. "Woodward: Tenet Told Bush WMD Case a Slam Dunk," *CNN*, April 19, 2004.

8. Jane Mayer, "The Manipulator, Ahmed Chalabi Pushed a Tainted Case for War," *New Yorker*, June 7, 2004.

9. Interview with Jay Garner on March 14, 2016, by Gary Vogler.

10. John Dizard, "How Ahmed Chalabi Conned the Neocons," *Salon*, May 4, 2004.

11. Daniel Ammann, *The King of Oil: The Secret Lives of Marc Rich* (New York: St Martin's, 2009), 152.

12. Interview with Mohammed Al Jibouri on June 7, 2016, by Gary Vogler.

13. David Sanger and Thom Shanker, "A Nation at War: White House; Bush Says Regime in IRAQ Is No More; SYRIA Is Penalized," *New York Times*, April 16, 2003.

14. Gary Vogler, *Lessons Learned—The Iraq Energy Sector* (Fairfax: Howitzer, 2016), appendix.

15. Ed Vuillamy, "Israel Seeks Pipeline for Iraqi Oil," *Guardian*, April 19, 2003.

16. Steve Coll, *Private Empire: ExxonMobil and American Power* (New York: Penguin, 2012), 221.

17. Jack Herman, "A Whole New Ballgame Overseas," *St. Louis Post*, February 20, 1989.

18. Follow-up interview with Michael Mobbs, December 10, 2015, by Gary Vogler.

19. Dan Eggen and Jerry Markon, "2 Senior AIPAC Employees Ousted," *Washington Post*, April 21, 2005.

20. David L. Phillips, *Losing Iraq: Inside the Postwar Reconstruction Fiasco* (New York: Basic, 2005), 71.

21. "JINSA Welcomes Dr. Michael Makovsky as Its New Chief Executive Officer and Executive Director," *JINSA*, April 18, 2013, http://jinsa.org/publications/press-releases/jinsa-welcomes-dr-michael-makovsky-its-new-chief-executive-officer-and-e.

22. Vuillamy, "Israel Seeks Pipeline."

23. Amiram Cohen, "US Checking Possibility of Pumping Oil from Northern Iraq to Haifa, via Jordan," *Haaretz*, August 25, 2003.

24. Stephen Scheer, "Netanyahu Says Iraq-Israel Oil Line Is No Pipe-Dream," *Reuters*, June 20, 2003.

25. Steve Sailer, "The Scooter Libby-Marc Rich Connection," *iSteve Blog*, October 31, 2005.

26. Ammann, *The King of Oil*, 64.

27. Ibid., 184.
28. Ibid., 152.
29. Ibid., 242.
30. Ibid., 137.
31. Ibid., 208.
32. Charles Grant, "Blair's Five Wars," *Prospect*, October 30, 2003.
33. Ammann, *The King of Oil*, 152.
34. Vuillamy, "Israel Seeks Pipeline."
35. National Security Archive, no. 82, edited by Joyce Battle, February 25, 2003, Documents 31, 48. Joyce Battle, *Shaking Hands with Saddam Hussein: The U.S. Tilts toward Iraq, 1980–1984*, National Security Archive Electronic Briefing Book No. 82, February 25, 2003, http://nsarchive.gwu.edu/NSAEBB/NSAEBB82/.
36. Christison and Christison, "The Bush Neocons and Israel."
37. Kelly and Ingersoll, "The Staggering Cost."

Epilogue

1. *Iraq Energy Institute*, "Exclusive Interview with H. E. Thamir Al-Ghadhban," December 14, 2015.
2. Ben Lando, "Occidental Plans Iraq Exit," *Iraq Oil Report*, November 10, 2015.
3. Ben Lando, "Exxon Splits West Qurna 1 Stake with CNPC and Pertamina," *Iraq Oil Report*, November 28, 2013.
4. *Iraq Energy Institute*, "Exclusive Interview with H. E. Thamir Al-Ghadhban."
5. Ibid.
6. Ben Lando, "KRG Tanker Unloads in Israel after Abandoning US Delivery," *Iraq Oil Report*, February 27, 2015.
7. Ben Lando and Patrick Osgood, "Court Blocks KRG's Oil Route to US as Israel Exports Grow," *Iraq Oil Report*, September 24, 2015.
8. Ibid.
9. *Iraq Energy Institute*, "Exclusive Interview with H. E. Thamir Al-Ghadhban."

Bibliography

Personal Documents

Workbooks and date books used during 2002 to 2011: All of the author's daily notebooks have been maintained and used as reference material. These books cover the deployment period of March 2003 through 2011. The prewar work at the Pentagon was in a classified environment, so daily notes were much more limited. A personal calendar for the period of October 2002 through March 2003 was used to cover this classified period.

A partial list of unclassified emails sent and received during 2003: Emails in Kuwait and Iraq during 2003 were limited to Yahoo and AOL accounts and eventually to a military unclassified account in Baghdad. Electronic files of unclassified emails for 2003 were maintained while in Kuwait and Baghdad. They were used as reference material.

Internet Subscriptions

Iraq Oil Report (IOR): This paid subscription service has been operating since 2008, when it was started by Ben Lando. Lando has grown this service by adding additional Western reporters and several Iraqi reporters. Today, it is considered one of the most respected services for the Iraq oil sector. It is used extensively in the book for information after the author's departure from Iraq in 2011.

Iraq Oil Forum: This online service is owned and operated by Ruba Husari. Husari has covered the Iraq oil sector longer than any other reporter. She covered the sector before 2003 and was one of the early reporters at the oil ministry in April 2003. Forum does not have the resources that IOR has developed, but Husari has the personal contacts and longtime relationships with key oil sector Iraqis that IOR does not have.

Interviews

Agoglia, John, Colonel, USA, Retired, November 18, 2014, Arlington, VA.
Callicotte, Harold, Lieutenant Colonel, USAR, Retired, March 19, 2015, telephone.
Carruthers, Robert, GRD Contractor, UK Citizen, December 19, 2015, telephone.
Cook, Christian, Major, USA, October 23, 2015, telephone.
Cripwell, Richard, Major General, UK Forces, October 29, 2015, Washington, DC.
DaSilva, Joseph, Major, USA, October 13, 2015, telephone.

Dudding, Wayne, Lieutenant Colonel, USAR, Retired, November 21, 2014, and March 1, 2015, Wheeling, WV.
Fritzley, Richard, Captain, USNR, Retired, October 3, 2015, telephone.
Garner, Jay, Lieutenant General, USA, Retired, March 14, 2016, telephone.
Lindsey, Stephen, DOS Civilian, US Embassy, Baghdad, December 4, 2015, telephone.
Maguire, Robert, Esq., August 18, 2015, Sterling, VA.
McKee, Robert, VP Conoco, Retired, March 23, 2015, Houston, TX.
McNinch, Hamish, Brigadier General, UK Forces, Retired, November 19, 2015, telephone.
Mobbs, Michael, DOD, Retired, March 13, 2015, Washington, DC; Follow-up phone call on December 10, 2015.
Obaidi, Dr. Ali, Director General, Ministry of Oil, Retired, December 17, 2014, and December 30, 2015, Sterling, VA.
O'Donnell, Thomas, Colonel, USA, Retired, March 22, 2015, Houston, TX.
Quanrud, Pamela, FSO, DOS, June 15, 2015, Washington, DC.
Ramos, Betty J., Captain, USAR, November 10, 2014, telephone.
Raphel, Robin, Ambassador, Retired, June 17, 2015, Washington, DC.
Rudd, Gordon W., Colonel, USA, Retired, June 12, 2016, Manassas, VA.
Shelton, Paul, Lieutenant Colonel, USMC, Retired, November 8, 2014, Vienna, VA.
Stinson, J. Michael, VP Conoco, Retired, March 23, 2015, Houston, TX.
Turner, Clarke, DOE, Retired, March 6, 2015, Washington, DC.

Books

Ahmad, Muhammad Idrees. *The Road to Iraq: The Making of a Neoconservative War.* Edinburgh: Edinburgh University Press, 2014.
Allawi, Ali A. *The Occupation of Iraq: Winning the War, Losing the Peace.* New Haven: Yale University Press, 2007.
Ammann, Daniel. *The King of Oil: The Secret Lives of Marc Rich.* New York: St. Martin's, 2009.
Bremer, L. Paul, III. *My Year in Iraq: The Struggle to Build a Future of Hope.* New York: Simon & Schuster, 2006.
Bush, George W. *Decision Points.* New York: Crown, 2010.
Chandrasekaran, Rajiv. *Imperial Life in the Emerald City: Inside Iraq's Green Zone.* New York: Knopf, 2006.
Cheney, Dick. *In My Time: A Personal and Political Memoir.* New York: Threshold, 2011.
Chernow, Ron. *Titan: The Life of John D. Rockefeller, Sr.* New York: Random House, 1998.
Clark, Wesley K. *Winning Modern Wars: Iraq, Terrorism, and the American Empire.* New York: Public Affairs, 2003.
Coll, Steve. *Private Empire: ExxonMobil and American Power.* New York: Penguin, 2012.

Diamond, Larry. *Squandered Victory: The American Occupation and the Bungled Effort to Bring Democracy to Iraq.* New York: Times, 2005.

Eberly, Don. *Liberate and Leave: Fatal Flaws in the Early Strategy for Postwar Iraq.* Minneapolis: Zenith, 2009.

Feith, Douglas J. *War and Decision: Inside the Pentagon at the Dawn of the War on Terrorism.* New York: HarperCollins, 2008.

Franks, Tommy, with Malcolm McConnell. *American Soldier: General Tommy Franks.* New York: Regan, 2004.

Friedman, Thomas L. *Hot, Flat, and Crowded: Why We Need a Green Revolution—and How It Can Renew America.* New York: Farrar, Straus and Giroux, 2008.

Galbraith, Peter W. *The End of Iraq: How American Incompetence Created a War without End.* New York: Simon & Schuster, 2006.

Gates, Robert M. *Duty: Memoirs of a Secretary at War.* New York: Knopf, 2014.

Gellman, Barton. *Angler: The Cheney Vice Presidency.* New York: Penguin, 2008.

Gordon, Michael R., and Bernard E. Trainor. *Cobra II: The Inside Story of the Invasion and Occupation of Iraq.* New York: Pantheon, 2006.

———. *The Endgame: The Inside Story of the Struggle for Iraq, from George W. Bush to Barack Obama.* New York: Vintage, 2012.

Graham, Bradley. *By His Own Rules: The Ambitions, Successes, and Ultimate Failures of Donald Rumsfeld.* New York: Public Affairs, 2009.

Haass, Richard N. *War of Necessity-War of Choice: A Memoir of Two Iraq Wars.* New York: Simon & Schuster, 2009.

Hoeffel, Joseph M. *The Iraq Lie: How The White House Sold the War.* San Diego: Progressive, 2014.

Hoffman, David E. *The Oligarchs: Wealth and Power in the New Russia.* New York: Public Affairs, 2002.

Maass, Peter. *Crude World: The Violent Twilight of Oil.* New York: Knopf, 2009.

Mann, James. *Rise of the Vulcans: The History of Bush's War Cabinet.* New York: Penguin, 2004.

Mansoor, Peter R. *Surge: My Journey with General David Petraeus and the Remaking of the Iraq War.* New Haven: Yale University Press, 2013.

Marr, Phebe. *The Modern History of Iraq.* Boulder: Westview, 1985.

Miller, T. Christian. *Blood Money: Wasted Billions, Lost Lives, and Corporate Greed in Iraq.* New York: Little, Brown, 2006.

Muttitt, Greg. *Fuel on the Fire: Oil and Politics in Occupied Iraq.* London: Bodley Head, 2011.

Olsen, Kim. *Iraq and Back: The Inside Wars to Win the Peace.* Annapolis: Naval Institute Press, 2006.

Packer, George. *The Assassin's Gate: America in Iraq.* New York: Farrar, Straus and Giroux, 2005.

Pelletiere, Stephen. *America's Oil Wars.* Westport: Praeger, 2004.

Phillips, David L. *Losing Iraq: Inside the Postwar Reconstruction Fiasco.* Boulder: Westview, 2005.

Ricks, Thomas E. *Fiasco: The American Military Adventure in Iraq*. New York: Penguin, 2006.

———. *The Generals: American Military Command from World War II to Today*. New York: Penguin, 2012.

Roston, Aram. *The Man Who Pushed America to War: The Extraordinary Life, Adventures, and Obsessions of Ahmed Chalabi*. New York: Nation, 2008.

Rudd, Gordon. *Reconstructing Iraq: Regime Change, Jay Garner, and the ORHA Story*. Lawrence: University Press of Kansas, 2011.

Rumsfeld, Donald. *Known and Unknown: A Memoir*. New York: Penguin, 2011.

Sanchez, Riccardo S., with Donald T. Philips. *Wiser in Battle: A Soldier's Story*. New York: HarperCollins, 2008.

Simmons, Matthew R. *Twilight in the Desert: The Coming Saudi Oil Shock and the World Economy*. Hoboken, NJ: Wiley, 2005.

Sky, Emma. *The Unraveling: High Hopes and Missed Opportunities in Iraq*. New York: Public Affairs, 2015.

Special Inspector General for Iraqi Reconstruction (SIGIR). *Hard Lessons: The Iraqi Reconstruction Experience*. Washington, DC: Government Printing Office, 2009.

Suskind, Ron. *The Price of Loyalty*. New York: Simon & Schuster, 2004.

Tarbell, Ida M. *The History of the Standard Oil Company*. New York: McClure, Phillips, 1904.

Tenet, George, with Bill Harlow. *At the Center of the Storm: My Years at the CIA*. New York: HarperCollins, 2007.

Vogler, Gary. *Lessons Learned—The Iraq Energy Sector*. Fairfax: Howitzer, 2016.

Woodward, Bob. *Plan of Attack*. New York: Simon & Schuster, 2004.

———. *State of Denial: Bush at War*. Part 3. New York: Simon & Schuster, 2006.

Yergin, Daniel. *The Prize: The Epic Quest for Oil, Money & Power*. New York: Free, 2008.

———. *The Quest: Energy, Security, and the Remaking of the Modern World*. New York: Penguin, 2011.

Index